EXPLORING
SPACE
A GUIDE TO
EXPLORATION OF THE UNIVERSE

by

MARK R. CHARTRAND, Ph.D.

Illustrated by
RON MILLER

GOLDEN PRESS · NEW YORK
Western Publishing Company, Inc.
Racine, Wisconsin

FOREWORD

The author would like to thank many people for their contributions to this book: first of all, Caroline Greenberg and Remo Cosentino at Golden Press for their tireless help, expertise, and friendship; colleagues Bonny Lee Michaelson, Dr. Robert E. Murphy of NASA, Frederick I. Ordway, III, Ian Pryke of ESA, Leonard David, Dr. Glen P. Wilson, and commercial astronaut Charles Walker for their assistance; other colleagues on the Board of Directors and the staff of the National Space Society; Ron Miller for his artistic contributions; and the people at the photo libraries of NASA Headquarters, Johnson Space Center, and Jet Propulsion Laboratory, as well as others who supplied photographs and assistance.

Finally, I would like to express my indebtedness to the late Dr. Wernher von Braun, whom I never quite had a chance to meet, but whose ideas and writings, many years ago, first interested me in space exploration.

M.R.C.

CONTENTS

INTRODUCTION

The space program is one of the most exciting and significant endeavors in human history. We have reached out in person to Earth orbits and the Moon, with our robot spaceprobes beyond the edge of the solar system, and with our instruments to the edge of the universe.

Although the Space Age began as a contest between the Soviet Union and the United States of America (and to some extent still is), many other nations are now "space-capable," including Britain, France, India, Japan, China, and Israel. Several more will gain that capability soon. Hundreds of operational satellites and spaceprobes are in orbit right now, and the number is growing. Moreover, not only nations but private firms are now building and launching both rockets and satellites.

Over the past couple of decades, the Soviet Union has typically launched about 100 satellites a year, the U.S. 20 to 25, and other nations combined another half dozen or so. Most of the Soviet launches are military in nature, although that nation makes less of a distinction between military and civilian spacecraft than do Western nations.

The main focus of this book is the civilian space effort around the world. Much space hardware and terminology, however, came from military programs, and so a section on military uses of space is also included.

Bootprint in the lunar dust
NASA

COSTS OF SPACE EXPLORATION seem high, but there are also great rewards. In the United States, the entire civilian space budget cost less than 8/10 of 1 percent of the federal budget yearly during the 1980s. A major space endeavor, such as the Galileo spaceprobe to Jupiter, costs each person in the U.S. only about $4.50, about the price of a hamburger, fries, and a shake in 1990.

BENEFITS OF SPACE EXPLORATION are great but are not easy to quantify. Besides the intellectual rewards of exploring and understanding our place in the universe, there are real economic benefits. What is the value of a human life saved by a search-and-rescue satellite? How much money is it worth to know a hurricane will hit a certain city? What is it worth to maintain a lead in high technology? Even more nebulous, how much is it worth to know about the satellites of Jupiter or the existence of black holes?

While it is difficult to come up with definite numbers, it is clear that space programs around the world have made it possible to manage better the planet that has been called "Spaceship Earth," and help make it a better, safer place to live.

Over the next decade commercial firms, not just governments, will begin to use the results of space programs to benefit everyone on Earth.

Earth photographed from space
NASA

OBSERVING ROCKETS AND SATELLITES can be exciting, but you have to be in the right place at the right time.

Rocket launches take place from only a few locations. In the U.S., the only one easily available to the public is the Kennedy Space Center in Florida. Although the public is not allowed on the grounds during a launch, there are many nearby beach areas from which launches can be seen. Civilian launches are usually announced in advance; military launches almost never are. Night launches are particularly spectacular. Once you have seen a real rocket launch you will never forget it.

Satellites are visible from Earth when they pass nearly over your location. Only a few are bright enough to be

Rocket launch   NASA

seen with the unaided eye. The best times are near dusk and dawn, when you are in darkness but the satellite high above you is still in sunlight. Some local planetariums, observatories, space-interest groups, and NASA facilities offer information about satellite passages visible from your area. Personal-computer programs are available for tracking satellites, particularly those satellites used by amateur radio operators.

SPACE LAW There is no formal legal definition of where airspace ends and "outer" space begins. There is, however, an operational definition that is more-or-less accepted: 60 miles up, the height at which a satellite can make at least one orbit before air drag makes it fall back to Earth.

Spacecraft are subject to laws very different from those that apply to the most nearly analogous situations: flying in the air or sailing on the high seas in international waters. Whereas there are admiralty laws dealing with flotsam, jetsam, and abandoned ships, no such laws apply in space. And while it is an accepted part of international law that a nation owns the earth below and the airspace over its borders, the same is not true of "outer" space.

The United States and most space-faring nations are parties to several space treaties. These deal with registering space objects, the rescue of astronauts, liability in case a spacecraft does damage on Earth, and peaceful uses of space. Another treaty, which the United States did *not* sign, claims that the Moon and other bodies in space are "the common heritage of mankind." Many space entrepreneurs are afraid this will be interpreted to mean that no one can mine the Moon, asteroids, or other planetary bodies.

Existing space treaties were written at a time when no one thought private companies could afford space activities, so the treaties do not recognize private enterprise. For instance, there does not exist in space the concept of a "common carrier." A nation that allows a launch, or whose citizens own the satellite, is liable for damages in case something goes wrong. This has had a dampening effect on private enterprise. It will be important to revise current space laws and make new ones that will carry us into the 21st century.

As space becomes a place to do business, new laws and regulations will be needed in such areas as patents, citizenship, taxation, and insurance.

The Battle of K'ai-fung-fu

THE FIRST ROCKETS

The principle of the reaction motor was known as far back as 360 B.C., when Aulus Gellius described a steam-powered model of a pigeon. A Greek named Hero is said to have invented a steam-driven rotating "aeropile" about 2,000 years ago.

Most people credit the Chinese with the invention of the rocket itself, powered by burning black powder, a mixture of charcoal, saltpeter, and sulfur. Some historians claim the Chinese had powder rockets almost 1,000 years ago, but it is certain that rocket-powered arrows were used by the Chinese military in the 13th century. "Firearrows" drove off the attacking Mongols at the Battle of K'ai-fung-fu in 1232. Other armies soon took up the idea, and in 1241 Mongol tribes used the same kind of weapons at the Battle of Sejo (near present-day Budapest).

Also in the 13th century, the English scientist Roger Bacon and the Arab scientist al-Hasan al-Rammah described black powder and rocketry. Soon this technology was in use by armies throughout Europe and Asia.

An aeropile

Wan Hoo's rocket-piled chair

A few non-military uses were tried, including fireworks. It is said that around the year 1500 the Chinese Wan Hoo had 47 black-powder rockets attached to a sedan chair and ignited at the same time by 47 servants. He was not seen again.

By the late 1700s and early 1800s, rockets carrying explosive warheads were a standard but small part of the arsenals of most countries.

Sir William Congreve greatly improved rockets for warfare. His rockets, carrying either incendiary or explosive warheads,

Congreve rockets

could be fired a couple of miles and weighed up to 60 pounds. They first saw major use in 1806. Congreve rockets were later used by the British in the War of 1812, notably in the night attack on Fort McHenry. These were immortalized in the words of Francis Scott Key when he wrote of "the rocket's red glare."

Later that century, Congreve's rockets became obsolete because of advances in artillery. Similar rockets were adapted for use in throwing lifesaving ropes to stranded ships, and to some extent this is still done today.

Around the beginning of the 20th century advances in technology, and popular fantasies of trips to the Moon and life on other worlds, led to the work of the three great rocket pioneers and visionaries: Konstantin Tsiolkovsky in Russia, Hermann Oberth in Germany, and Robert Goddard in the United States.

In the mid-twentieth century, rockets again entered the arsenals of most nations.

The Battle of Ft. McHenry

Spaceflight fantasies

SPACEFLIGHT in the imagination is thousands of years old. Science fiction, until the 20th century, was much more fiction than science.

Lucian of Samosata, a Greek of the 2nd century A.D., wrote of a trip to the Moon by a sailing ship caught in a whirlwind. In 1010, the Persian poet Firdusi described a throne pulled to the Moon by eagles. The astronomer Johannes Kepler wrote a fanciful tale of a lunar trip in 1634, and was perhaps the first to realize that the air did not extend all the way to the Moon. About the same time, Englishman Francis Godwin wrote of men carried to the Moon by geese. Around 1650 the famous writer Cyrano de Bergerac wrote stories of trips to the Moon and the Sun. His first plan of travel was to tie bottles of dew to his belt and thus rise into space when the Sun evaporated the dew! Another idea he had was to use fireworks rockets. Around 1705 Daniel Defoe, better known for *Robinson Crusoe*, wrote a tale of lunar travel.

Only in the late 19th century did such stories become more realistic. The father of modern science fiction is Jules Verne, who wrote *From the Earth to the Moon* in 1865. The concept of a manned space station and commercial navigation satellite comes from Edward Everett Hale and his 1869 story "The Brick Moon."

One of the earliest spaceflight motion pictures was Georges Melies' short film *A Trip to the Moon* in 1902. Most science-fiction films have been little more than fantasies or "space westerns." Only a few films have approached the subject realistically. The first such film was *The Woman in the Moon*, directed by Fritz Lang in 1929. His technical advisors went on to establish the early German rocket program. This film made a permanent contribution to the space program when Lang invented, purely for dramatic purposes, the countdown before launch. In 1950, producer George Pal made *Destination Moon*, a realistic depiction of what spaceflight might be like.

The spaceship controls in the 1939 film *Buck Rogers*   ORDWAY

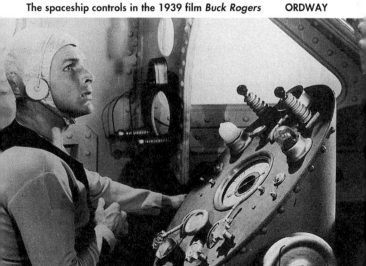

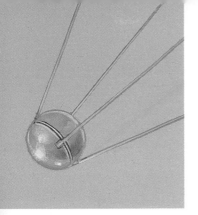

Sputnik 1

THE SPACE AGE BEGAN on October 4, 1957, when the Soviet Union launched the world's first artificial satellite, Sputnik 1. Though it weighed only 184 pounds, was less than 2 feet in diameter, and stayed in its 140-mile-high orbit only 92 days, its launch provoked shock waves of public reaction around the world.

This first satellite was launched from the Tyuratam (Baikonur) Cosmodrome on a Type-A rocket. It carried experiments to study the density of the atmosphere at its orbital altitude and the transmission of radio waves through the atmosphere. Its instruments were powered by batteries, which ran out three weeks after launch.

Sputnik 2, weighing 1,120 pounds and carrying a live dog, was launched on November 3, 1957.

The U.S. did not launch its first satellite, Explorer 1, until January 31, 1958, after initiating a crash program in an attempt to catch up with the Soviets, and after two attempted Vanguard satellite launches had failed. The first successful Vanguard satellite went into orbit on March 17, 1958. An Atlas rocket became the first communications satellite later that year.

Explorer 1 model   NASA

14

MANNED SPACEFLIGHT began on April 12, 1961, when Soviet Lt. Yuri Gagarin made one orbit lasting 108 minutes, 250 miles above Earth, in his Vostok-1 spacecraft.

The first American in space was Alan B. Shepard, who took a 115-mile-high suborbital flight in a Mercury-Redstone capsule on May 5, 1961. He was followed in another similar flight by Virgil I. (Gus) Grissom on July 21. On August 6, the Russian cosmonaut Gherman Titov spent 25 hours and 17 orbits in space aboard Vostok 2.

The first American to orbit the planet was John Glenn. On February 20, 1962, he rode a Mercury-Atlas rocket that

John Glenn   NASA   Alan Shepard   NASA

made three orbits 160 miles up in 5 hours. Scott Carpenter followed on May 24 in a similar mission.

The first woman in space was Valentina Tereshkova, aboard Vostok 6 on June 16, 1963. She made 48 orbits in 71 hours. No other woman went into space until 20 years later, when astronaut Sally Ride was part of the five-member crew aboard the space shuttle Challenger on June 18, 1983.

15

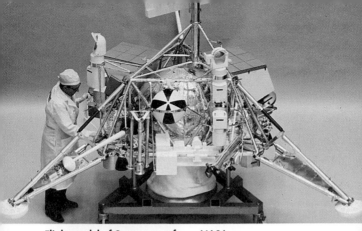

Flight model of Surveyor craft   NASA

LUNAR EXPLORATION began when the Soviet Lunik 1 flew within 4,600 miles of the Moon in January 1959. That March the U.S. Pioneer 4 passed the Moon at a distance of 37,000 miles. In September, Lunik 2 became the first man-made object to hit the Moon. Lunik 3, in October, was the first to take photographs of the far side of the Moon.

Beginning in 1964, U.S. Ranger craft crash-landed on the Moon, photographing areas that might be used for later manned landings. Lunik 9 in January 1966 became the first craft to soft-land on the Moon, returning television pictures of the surface.

U.S. Surveyor craft also soft-landed, taking photographs and digging trenches to analyze surface properties. Lunar Orbiters provided excellent photographs of most of the surface. Several orbiting Soviet Lunik craft also mapped the Moon. The Soviets sent several unmanned craft to the Moon to scoop up soil and return it to Earth. Some deployed a small roving vehicle called Lunokhod to sample lunar soil over a wide area.

FIRST VISIT TO ANOTHER WORLD was the goal of the U.S. Apollo program. A three-man spacecraft, consisting of a command module attached to a two-section lunar module, established orbit around the Moon. The lunar module carried two men to the surface, and its separable ascent stage brought them back to lunar orbit to rendezvous with the command module, which then returned to Earth.

The first manned lunar orbit was achieved by Apollo 8, launched December 21, 1968, carrying Frank Borman, James Lovell, and William Anders. Apollo 11, launched July 16, 1969, made the first manned lunar landing. Michael Collins remained in the command module while Neil Armstrong and Edwin Aldrin landed on the Moon's Sea of Tranquillity at 4:17 P.M EDT on July 20. Neil Armstrong became the first human to set foot on another world at 10:56 P.M.

Five more Apollo missions landed on the Moon. Apollo 13 encountered problems and returned to Earth after swinging around the Moon. The successful missions did experiments and returned almost 850 pounds of lunar rocks for analysis. The last mission was Apollo 17, in December 1972. No one has been to the Moon since.

Apollo 11 on the Moon   NASA

THE FIRST SPACE STATION was the Soviet Salyut 1, launched April 19, 1971. Weighing over 25 tons, it had a total length of 47 feet and a maximum diameter of 13 feet. A crew of three were carried up to the station by Soyuz vehicles. Unfortunately, their Soyuz-11 craft malfunctioned as the crew returned to Earth, and the three cosmonauts died. The Salyut-1 station was shortly thereafter commanded to reenter the atmosphere and burn up after only six months in orbit.

Other Salyut stations followed. Salyut 6, launched in 1977, was a big step in Soviet spaceflight. Over 49 feet long, weighing 41,000 pounds, and with docking ports for two Soyuz craft, this space station was occupied by 33 different crewmen and docked with 30 different spacecraft during its 3-year-and-8-month lifetime. Unmanned Progress craft repeatedly carried supplies to the crews.

Salyut 7, launched April 19, 1982, was approximately the same size as its predecessor, but it had many improvements. Its crews have included the first woman to "walk" in space, Svetlana Savitskaya. It fell back to Earth in 1991.

Saylut 1

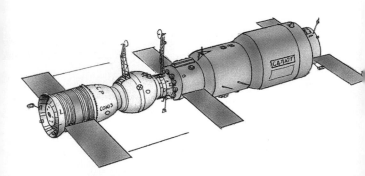

First U.S. space station   NASA   Inside Skylab   NASA

SKYLAB, in 1973, was the first and so far the only American space station. It was constructed inside the upper stage of a Saturn rocket, with two main rooms for experiments and living quarters. The weight of this entire 118-foot-long, 21-foot-diameter station was almost 200,000 pounds. The first 28-day mission was followed by another a month later, when a three-man crew spent 59 days in space. The third and last crew broke space-endurance records up until then (1974) by staying 84 days in orbit. The crews performed many important experiments in solar studies, space medicine, materials processing, and Earth studies.

Atmospheric drag slowly pulled Skylab lower, and Congress cut funds that would have provided a booster engine to keep the station in orbit longer. Finally, on July 11, 1979, Skylab fell to Earth over the Indian Ocean. A few pieces landed in Australia without damage to property. Since then America has had no space station, and will not have one until later this decade.

SPACEPROBES to the planets began in 1959, when Lunik 1 became the first man-made craft to escape Earth permanently. The first close-up pictures of Mercury and Venus came from the American Mariner-10 probe in 1973. The U.S.S.R. also has many exploratory firsts for Venus: first flyby, Venera 1 in 1961; first Venusian impact, Venera 3 in 1965; first Venus atmosphere probe, Venera 4 in 1967; first soil analysis, Venera 8 in 1972; first pictures from the surface, Venera 9 in 1975.

The Soviet Union also boasts the first Mars flyby, by Mars 1 in 1962; first Mars crash landing, Mars 2 in 1971; and the first soft landing on Mars, Mars 3 the same year. First pictures of Mars from space came from the U.S. Mariner 4 in 1965, and the first photographs from the surface were made by the two U.S. Viking landers in 1976.

The first close-up pictures of Jupiter and its moons came from the U.S. Pioneer-10 probe in 1972, and of Saturn and its satellites by Pioneer 11 in 1973. Voyager-1 and Voyager-2 probes took more detailed pictures. Voyager 2 went on to fly by Uranus in 1986, and by Neptune in 1989.

The first mission to a comet was by the U.S. International Cometary Explorer (ICE) to Comet Giacobini-Zinner in 1985. In 1986, the U.S.S.R., the European Space Agency, and the Japanese all sent missions to rendezvous with Comet Halley. This decade spaceprobes will encounter asteroids, study Jupiter's atmosphere, and study the Sun's poles.

Venera 9   TASS   Venus' surface seen from Venera   TASS

Mariner 10 (left); Venus (middle) and Mercury's surface (right) seen by Mariner 10   NASA

Voyager spacecraft (left); Jupiter (middle) and Saturn (right) seen by Voyager   NASA

Above: Viking lander; Mars' surface seen by Viking   NASA

Below: Giotto comet probe (left)   ESA; Comet Halley as seen by Giotto (right)   MPAE

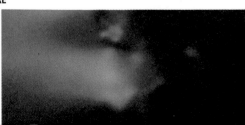

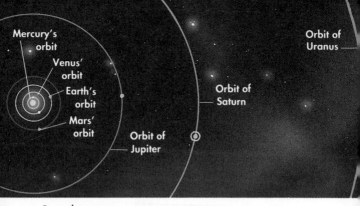

Mercury's orbit

Venus' orbit

Earth's orbit

Mars' orbit

Orbit of Jupiter

Orbit of Saturn

Orbit of Uranus

Our solar system

THE SOLAR SYSTEM

The solar system is the region in which all our spacecraft, so far, have explored. While our earthly and spaceborne telescopes have probed to the depths of the universe, our craft have gone only just beyond the edge of the planetary system.

The Sun is the center of the solar system, and has more than 99 percent of all its material. Orbiting the Sun are nine major planets, more than five dozen satellites, hundreds of thousands of minor planets (asteroids), and perhaps billions of comets. Except for the comets and some of the asteroids, these all revolve in paths limited to a narrow disk around the Sun. Comets may have highly inclined orbits. Most are located many billions of miles from the Sun, only occasionally plunging into the inner solar system where we can see them.

Astronomers measure planetary distances in terms of the Earth's distance from the Sun, called an astronomical unit (a.u.), equal to about 93 million miles. Pluto on average is more than 39 times this far from the Sun, but because of Pluto's

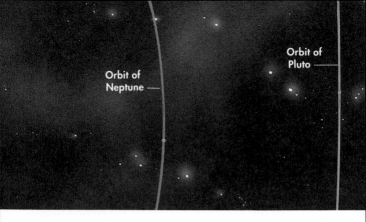

Orbit of
Neptune

Orbit of
Pluto

elliptical orbit, Neptune will actually be the most distant planet until March of 1999.

Another way to measure distance is the time it takes light or radio waves, moving at 186,000 miles per second, to travel that distance. Light takes about 8 minutes to go 1 a.u.; it takes 5 hours to get from the Sun to Pluto. The nearest star to our solar system is more than 4 light-years away, about 25 trillion miles.

Spaceprobes have robotically explored all the planets except Pluto, although Pioneer 10 has passed the orbit of Pluto and is now more than 4.2 billion miles away, the most distant man-made object. The limit of manned exploration, so far, is the Moon, only 240,000 miles from Earth.

In planning for space missions, a more important factor than time or distance is the energy needed to get somewhere. Each planet has a strong gravitational pull. A spacecraft going to another planet must first escape the gravitational field of Earth, travel to the planet, and then maneuver within *its* gravitational field. Along the way, the craft is subject to the gravity of the Sun.

23

EARTH'S ATMOSPHERE is composed of about 78 percent nitrogen, 21 percent oxygen, and 1 percent trace gases such as water vapor, carbon dioxide, and argon.

THE TROPOSPHERE, the lowest 10 miles of air, is where all weather occurs and all life exists. At the surface the pressure is almost 15 pounds per square inch. From the surface upward the pressure and density of the air lessen rapidly. Only a few aircraft can fly higher than 10 miles up, but the air here is still much too thick for an unpowered satellite to orbit. Yet at this altitude, the air is too thin for a person to live—except in a pressure suit or a pressurized cabin.

THE STRATOSPHERE lies from 10 miles up to about 30 miles. A very few aircraft can fly this high. Higher still is the *mesosphere,* extending up to about 50 miles, and above that the *thermosphere.* At 60 miles above Earth the pressure is only a billionth, and the density is only three ten-millionths, what it is at the surface.

THE EXOSPHERE is the name sometimes given to the outer-most tenuous atmosphere where it gradually blends into the vacuum of space.

There is no legal or physical "top of the atmosphere." But many people consider 60 miles up a practical limit, for at this altitude a small satellite can make one orbit. Most satellites orbiting Earth are more than 100 miles up, but even there the atmosphere exerts a resisting "drag" force which slows them down so that an orbit eventually "decays" and the satellite falls back to Earth. Usually it burns up like a meteor, but a few pieces of satellites have landed on Earth. The smaller, denser, and higher a satellite, the longer it will stay in orbit. Above a few thousand miles, a satellite experiences almost no atmospheric drag.

Structure of the atmosphere

Satellite

Exosphere

Typical
shuttle orbit

Aurora

Meteors

Stratosphere

Troposphere

THE SPACE ENVIRONMENT

Space is a harsh environment. Our spacecraft designers must take into account these new conditions, which differ greatly from those on Earth.

THE VACUUM OF SPACE causes many otherwise stable materials, such as rubber and many lubricants, to turn into gas and disperse. Such materials can only be used in satellites if they are enclosed. Materials like glass become more brittle in a vacuum. Humans must either work inside pressurized containers or use spacesuits for work outside.

WEIGHTLESSNESS is the condition of anything in space that is not under power; it is also called free fall or "microgravity" or zero g, where the "g" stands for "gravity." An

Satellite after launch from shuttle   NASA

object still has mass in space, even if it does not have any weight, so it still takes work to start it moving and to stop it. The lack of gravity creates complications in designing a spacecraft and in living in space.

SUNLIGHT strikes spacecraft, producing heat. Above the absorbing atmosphere the Sun is stronger, and exposed surfaces may get hotter than the temperature of boiling water. The lack of air means there is no convection to carry excess heat away from the satellite; the only way a satellite can get rid of heat is by radiation. On the other hand, when a satellite is in shadow its temperature may fall to more than 100 degrees below zero Fahrenheit in just a few minutes. Since temperature extremes can damage spacecraft components, thermal control is an important part of satellite design.

Simulating zero g in a watertank   NASA

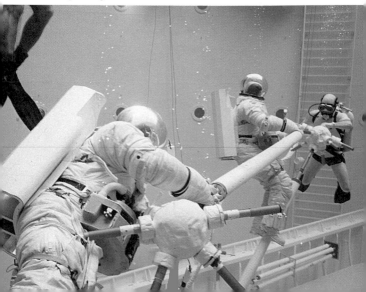

COSMIC RAYS are charged atomic particles, many of them the nuclei of atoms. These come from the Sun as well as from the depths of space. They may give the spacecraft an electrical charge, which can be damaging if it causes sparks. They also gradually reduce the efficiency of the solar cells (which convert sunlight into electrical energy). Very high-energy cosmic rays may even penetrate the electronic chips in the satellite's control circuitry, causing temporary or permanent changes in the commands given to the craft.

METEOROIDS are solid materials flying through space at tremendous speeds, usually many miles per second. Most meteoroids are smaller than a grain of sand, and are called micrometeoroids. Over a typical 10- to 15-year lifetime, a satellite will get a few pinpricks a tenth of an inch in size and a light "sanding" of its exposed surfaces, including the

Grabbing a satellite for repair   NASA

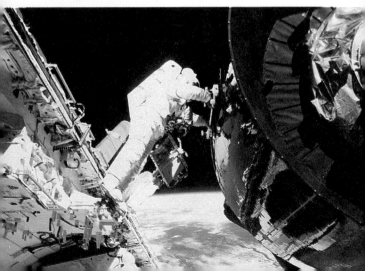

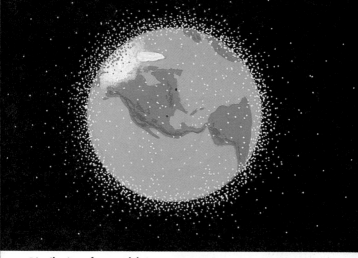

Distribution of space debris

solar panels. So far we know of no large (meaning greater than a few tenths of an inch in diameter) meteoroids having hit satellites. But the chances increase with time and with the size and greater number of spacecraft we have in space. Spacecraft have been hit by man-made space debris, however.

SPACE DEBRIS in Earth orbit is a serious and increasing problem. There are now well over 7,000 trackable objects in orbit. Only a few hundred are active satellites. There are more than 35,000 objects the size of a marble, perhaps millions of smaller pieces. The debris consists of used third stages of rockets, the remains of exploded (accidentally or intentionally) spacecraft, chips of paint knocked off satellites, hardware like bolts and springs released when satellites are

29

deployed, and parts knocked off satellites by collisions. Much of the debris has come from the testing of antisatellite weapons.

Even very small objects are hazardous because they are moving at speeds up to 25,000 miles per hour. One of the space shuttles was hit on its windshield by a tiny chip of aluminum paint, causing a small pit in the glass. Several satellites are strongly suspected of having been destroyed by collisions. The Soviet space station Salyut 6 had some of its external parts damaged by collision. The more satellites there are in space, and the more frequently they collide, producing still more fragments, the worse the problem will become. Not until after the year 2000 will we begin to have the technology to retrieve used-up satellites and bring them back from high orbits.

Space debris is becoming a problem for astronomers, too. Increasingly, debris shows up in photographs of the sky made with large telescopes, prompting false discoveries. One reported discovery of a pulsar turned out to be sunlight reflecting off the solar cells of a dead satellite!

Debris caused this pit in shuttle window   NASA

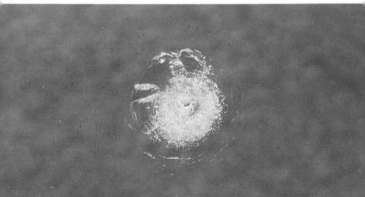

SPACE MEDICINE

When a human enters space there are physiological and psychological changes. Some body functions continue to alter as long as she or he is weightless; others reach a steady level within days or weeks. Upon return to Earth most body functions return to normal. Since the longest continuous period a cosmonaut has spent in weightlessness is about a year, we still do not know if some of these changes might become permanent after a very long stay in zero g.

SPACE SICKNESS, called "space adaptation syndrome" by NASA, affects about half of all people who go into space. Like other forms of motion sickness, it arises in the astronaut's inner

Astronaut Don Williams exercises aboard the space shuttle NASA

ear, the mechanism that senses orientation and acceleration. The symptoms include cold sweating, nausea, and vomiting. These could be dangerous to an astronaut in a spacesuit. Most people who get spacesick get over it in a couple of days. A few astronauts have briefly reexperienced it after their return to Earth.

CHANGES IN THE BODY include a loss of body fluids and solid material. In space, an astronaut's bones are no longer needed to support the body and so begin to weaken. The heart pumps faster; it grows larger but pumps less blood. Muscles grow weaker since they have less work to do. The number of red cells in the blood decreases while the number of white cells increases. After return to Earth it takes from weeks to months to bring blood cells back to pre-flight levels. Astronauts regularly exercise while in space to try to minimize these problems.

RADIATION is another problem in space. While not too serious in low Earth orbits, the high-energy particles trapped in the Van Allen belts around Earth are a danger. One belt extends from about 300 miles to about 750 miles up, the other from 6,000 miles to a height dependent on the activity of the Sun. Some radiation leaks through to lower altitudes.

Van Allen radiation belts

A medical checkup in space
NASA

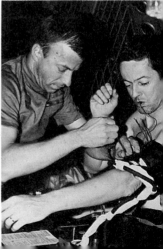

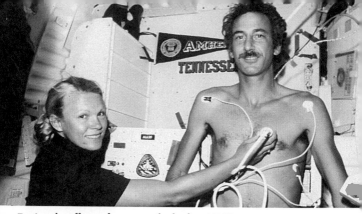

Testing the effects of space on the body   NASA

Astronauts heading away from Earth must pass quickly through these radiation zones to minimize exposure. Outside the Van Allen belts, space travelers are exposed to a continual dosage of radiation. Long-duration space missions must carry heavy shielding of some sort. Once in a while the Sun produces a solar flare that could give a lethal dose of radiation to an unprotected astronaut.

PSYCHOLOGICAL PROBLEMS come from lengthy confinement, living in an unnatural environment, close quarters, lack of privacy, and the ever-present hazards of living in space. Using studies of similar situations—for instance, aboard submarines—space-mission planners try to minimize potential problems. It will be important to provide many forms of recreation and relaxation for crews on long missions aboard space stations and manned missions to other planets.

MUCH RESEARCH is needed to determine human tolerances for extended missions either in low orbits or to the planets. This will be one of the more important goals of the projected U.S. space station later this decade.

33

LIVING IN SPACE

Weightlessness complicates even simple tasks in space. Spacecraft need special systems to provide the things required for life.

OXYGEN for breathing is carried in tanks. Some spacecraft use a reduced atmospheric pressure enriched in oxygen. Others have an environment much like natural air. Exhaled carbon dioxide is absorbed by chemical filters and the purified air is recirculated. Air must be in motion at all times in a spacecraft: an unmoving astronaut could suffocate in a bubble of her own exhaled breath because, without air circulation, it would not move away from her body.

EATING in space can be messy. Free liquid would float out of an open glass and could be a hazard if it escaped. Thus drinks come in squeeze bottles. Foods may be solids and pastes. Sauces are often used to make the foods stick to plates. Meals are usually prepared on Earth and stored for final preparation as needed.

WATER for drinking, cooking, and washing is carried in tanks, may be produced on board by fuel cells, and may be recycled from body waste. Excess water is also filtered out of the air.

Eating in space can be a challenge   NASA

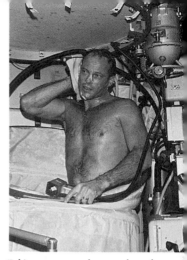

The space shuttle's galley area
NASA

Taking a space shower aboard
Skylab   NASA

Sally Ride sleeps aboard the
space shuttle   NASA

The toilet of the space shuttle
NASA

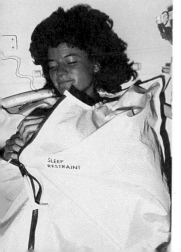

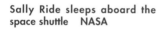

BODY-WASTE ELIMINATION is more complicated. In early spacecraft (and still in spacesuits) urine was collected by a tube and held in a bag. Solid waste was eliminated into a plastic bag held to the body with adhesive, and then stored until the end of the mission. The space shuttle, and the future space station, have a specially designed zero-g toilet. This uses air to pull the waste material into the toilet where a rotating fan separates the solids from the liquids. The liquid may be either sterilized and recycled or expelled from the spacecraft. Solid waste is dried and stored for return to Earth.

IN WEIGHTLESSNESS, as in a pool of water, the relaxed body naturally assumes a slightly curled-up position, with arms and legs floating freely in front of the body. An astronaut cannot lie on a bed to sleep; instead he zips himself into something much like a sleeping bag, which may be attached to a wall or ceiling or someplace else out of the way.

ORIENTATION can be a problem. The terms "wall" and "ceiling" have much less meaning in zero g. In the space shuttle the surface that is the floor when the shuttle is on the ground is still often thought of as "down" within the spacecraft. When thinking about things outside a spacecraft in orbit around a planet, it is most natural to think of "down" as the direction toward the planet.

Future space stations and space settlements will probably include growing plants, and possibly animals, as part of the ecosystem. Some types of algae digest waste products, and others produce oxygen. Large space structures may revolve to provide an "artificial gravity" that will eliminate many of the problems caused by weightlessness (although it will introduce some problems of its own).

SPACESUITS AND EVA

Spacesuits are designed to provide the astronaut with a self-contained environment for several hours of work outside the pressurized cabin. This is called extravehicular activity, or EVA. The suits supply oxygen, absorb exhaled carbon dioxide, and provide for some body-waste elimination. Some provide water through a tube in the suit's helmet. Sensors inside the helmet give the astronaut information about the suit's status, such as the amount of remaining oxygen.

Spacesuit and Manned Maneu-vering Unit

Newer suits allow full atmospheric pressure, rather than the reduced pressure used in older ones. The suits are in several parts: the lower part, or legs, attached to the upper part with airtight seals, and the helmet. The spacesuited astronaut may attach himself to a large backpack unit containing oxygen, or be tethered to the spacecraft by an umbilical hose. Once outside the cabin, astronauts move around by pulling and pushing themselves on the spacecraft.

NEWER SPACESUITS, lighter and more flexible than present ones, are being designed to make it easier for astronauts to work in space for extended periods of time.

37

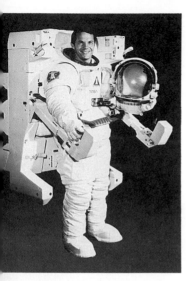

Astronaut uses a Manned Maneuvering Unit   NASA

MANNED MANEUVERING UNITS (MMUs) are jet-powered backpacks for excursions away from the craft. A hand controller much like one on a video arcade game is used to fire the jets in any combination of six directions, allowing motion or turning in any direction. Astronauts using MMUs don't have to be tethered to the spacecraft. Should an astronaut get into trouble, the spacecraft can always go after him.

One complication of zero g is that special tools must often be used because if, say, an astronaut tries to turn a screw, she will also turn, obeying Newton's Third Law (see p. 41). She must either brace herself against the craft, or use so-called "reactionless" tools.

One way to minimize this problem is to attach the astronaut firmly to part of the spacecraft, usually by clamps on the boots. For EVA on the space shuttle, astronauts are often attached to the Remote Manipulator Arm (see p. 67).

WALKING ON THE MOON, and in the future on other small planets, can present other problems. For instance, the Moon's gravity is only a sixth of Earth's, and the Apollo astronauts found that a bouncing, loping kind of gait was the best way to get around there.

ASTRONAUTS

The United States and many other nations call space travelers "astronauts," the Soviet Union calls theirs "cosmonauts," and the French term is "spationaut."

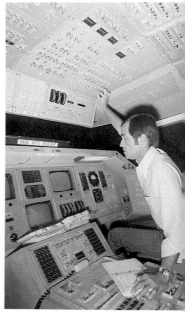

Training for a spaceflight in a simulator NASA

Most astronauts have come from the military services and have had training as test pilots. The first astronauts had little control over their craft. Later, especially in the space shuttle, the pilots have actually controlled the spacecraft much of the time. In the mid-1960s, the U.S. began recruiting non-pilot scientists and engineers from universities and industry for space missions.

For the space shuttle the crew consists of two pilots and up to five Mission Specialists and Payload Specialists. These latter have responsibility for deploying satellites, conducting experiments, and other non-piloting tasks. Most of the crew of the future space station will not be pilots.

All astronauts go through long and rigorous training. They must understand everything about their spacecraft, its mission, its payload (cargo), and what to do in case of emergencies. This is in addition to their particular areas of expertise, such as medicine, physics, astronomy, materials processing, and so on.

A payload specialist at work   NASA

The best preparation for being an astronaut, say the current astronauts, is becoming good in whatever field you like. Later you can apply that expertise to space.

NASA periodically advertises for astronauts to fill vacancies. Applicants are given several aptitude, psychological, and physical examinations. Those that are accepted become astronaut-candidates, and then finally astronauts when they have completed about a year of training. They get their astronaut wings upon completion of their first spaceflight. (The U.S. awards astronaut wings to anyone who has been more than 50 miles up; thus several pilots of the X-15 research plane are technically astronauts although they have not flown in spacecraft.) There are usually 50 to 100 men and women in the U.S. astronaut corps at any one time.

WHAT KEEPS A SATELLITE UP?

The motion of any object in space, whether that object is a micrometeoroid or the Sun, is controlled by the laws of physics.

NEWTON'S LAWS OF MOTION describe how an object moves:

I. A body at rest remains at rest, and a body in motion remains in motion, unless acted on by an outside force.

II. The change in the motion of a body is in the direction of, and proportional to, the strength of the force applied to it.

III. For every action, there is an equal and opposite reaction.

THE LAW OF GRAVITY describes the major force acting in space:

The gravitational attraction between two objects pulls them together with a force that is proportional to the product of their masses and inversely proportional to the square of their separation.

The Law of Gravity

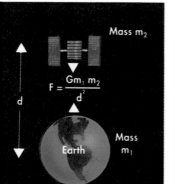

$$F = \frac{G m_1 m_2}{d^2}$$

Mass m_2

d

Earth

Mass m_1

An orbit is a balancing of forces

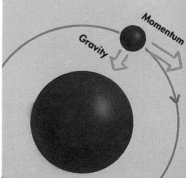

Momentum

Gravity

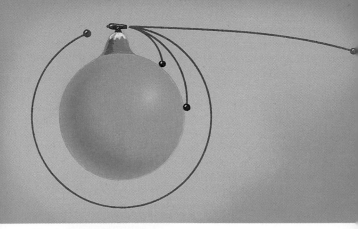

Newton's imaginary cannon

AN ORBIT is the motion of one object about another. An object remains in orbit because the momentum of its forward motion balances the gravitational pull between it and the other object. It is similar to the way you can swing a ball on the end of a string: the inward pull of the string balances the outward momentum of the ball. A satellite once placed in orbit (outside the atmosphere, which causes drag on aircraft and low-orbit spacecraft) can continue in that orbit without having to continually fire its rocket engines.

Newton gave the following example: suppose there were no atmosphere around the Earth to slow things down. A cannon on top of a very high mountain could fire a ball that goes some distance. As the ball travels, it falls toward the ground, but the curvature of Earth makes the ground "fall away" under it. If the cannon fires with sufficient energy, the ball can be made to travel at such a speed that the Earth falls away at the same rate that the ball falls toward Earth. In other words, the ball will never touch the ground: it will be in orbit.

ESCAPE VELOCITY is the minimum velocity needed to escape permanently from a planet. The speed required depends on the mass of the planet and the distance of the satellite from it. At any distance it will always be 1.4 times the speed necessary for a circular orbit.

KEPLER'S LAWS, discovered around 1610 by Johannes Kepler, describe any orbital motion:

   I. The orbit of each planet is an ellipse, with the Sun at one focus of the ellipse.

   II. There is a unique relationship between the distance of a planet from the Sun and how fast it moves: its speed will be slower when far from the Sun and faster when nearer the Sun.

   III. The average distance between the Sun and a planet, a, is related to the time it takes the planet to orbit the Sun once, t. Expressed mathematically, it is $t^2 = a^3$.

   If you substitute the word "satellite" for "planet" and "Earth" for "Sun," the laws are still true.

Kepler's Laws describe orbits

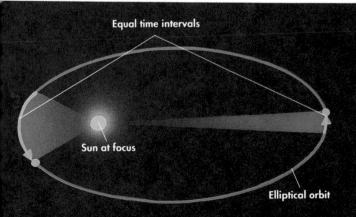

Equal time intervals

Sun at focus

Elliptical orbit

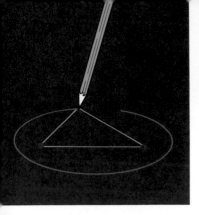

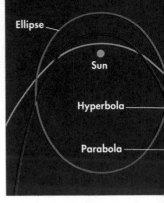

How to draw an ellipse · Types of open and closed orbits

AN ELLIPSE is a closed curve that can be made by sticking two pins into cardboard, looping a string around them, and then, keeping the string taut with the tip of a pencil, moving around the pins. The pencil will draw an ellipse. The points with the pins are called the foci (singular, focus) of the ellipse. The farther apart they are, the more eccentric the ellipse will be. If the pins are moved together, the ellipse becomes a circle.

The Sun would be where one of the pins is; the other focus is vacant. The long dimension of the ellipse is called the major axis; half of it is the semimajor axis, which is the average distance between the Sun and a planet. Circular and elliptical orbits are called "closed" orbits. If a satellite has reached escape velocity, its orbit will have the shape of a parabola or hyperbola, which are examples of "open orbits" since they don't repeat.

Actually, no orbit is perfectly elliptical because the gravitational pulls of all the other objects in the solar system also affect the motion. These smaller forces are called perturbations, and can be very complicated.

44

ORBITAL PARAMETERS describe an orbit in space. The size is described by the semimajor axis (half the long axis of the ellipse), which is the average distance from the center of Earth.

ECCENTRICITY describes the shape of the orbit. For closed orbits this is a number between 0 (for a perfectly circular orbit) and 1 (a parabolic escape orbit). A hyperbolic escape orbit has eccentricity greater than 1.

INCLINATION is the angle the orbit makes with the equator. An orbit lying over the equator has an inclination of 0 degrees; for one flying over the poles the inclination is 90 degrees. A satellite with inclination between 90 degrees and 180 degrees is going from east to west, and is said to have a retrograde orbit. Several other numbers describe the orientation of the orbit in space.

PERIOD, the length of time it takes to orbit once, depends on the altitude of the satellite. The space shuttle, orbiting a couple hundred miles high, has an orbital period of about 90 minutes.

Elliptical orbits come in many shapes

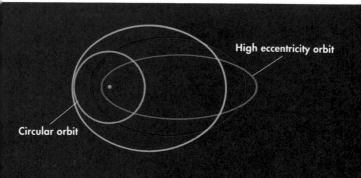

High eccentricity orbit

Circular orbit

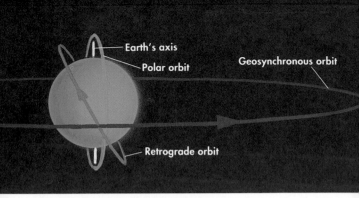

Inclination of an orbit to the equator

PERIGEE is the point on the orbit where the satellite is closest to the Earth (from "peri-" meaning "close" and "geo" meaning "Earth"). A satellite moves fastest at perigee.

APOGEE is the point in an orbit farthest from the Earth. Here a satellite moves slowest.

TRAJECTORY is a term sometimes used to describe a path from one orbit or planet to another. "Transfer orbit" is another term for the same thing. To go from one orbit to another one, you must fire a rocket to speed up the satellite (to get to a higher orbit) or slow it down (to get to a lower one). This will place the satellite in a transfer orbit. When it gets to the new orbit, you must again fire a rocket to give the satellite the correct speed for the new orbit.

LOW EARTH ORBITS are those within a few hundred miles of the surface of Earth. The lowest practical orbit is about 100 miles up. -

HIGHER ORBITS are those above about 1,000 miles. The higher the orbit, the longer the period (the time it takes to complete an orbit), and more of the Earth's surface that can be seen from the satellite.

GEOSYNCHRONOUS EARTH ORBIT is a special orbit 22,300 miles above Earth's surface. Here a satellite has a period of 24 hours, the same time it takes the Earth to rotate once.

GEOSTATIONARY ORBIT is a geosynchronous orbit with zero inclination. In this special case the satellite will seem to be stationary in the sky as seen from Earth. An antenna on Earth pointed at such a satellite would not have to move at all. The usefulness of such an orbit for communications satellites (commsats) was pointed out in 1945 by author-engineer Arthur C. Clarke. This orbit is often called a "Clarke orbit" in his honor.

SUN-SYNCHRONOUS ORBITS cause the satellite to pass over every place on Earth at the same local time of day. This is done by choosing the proper inclination for the orbit. A satellite in low orbit will have a Sun-synchronous path if its inclination is about 98 degrees.

Several types of orbits

Relationship of distance and period of orbits

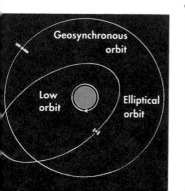

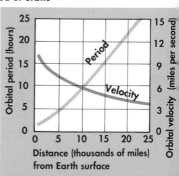

INTERPLANETARY ORBITS are those around the Sun. The numbers that describe these paths are similar to those of Earth orbits, but distance is measured from the Sun, and the inclination is measured compared to the Earth's orbit around the Sun. The nearest and farthest points in orbit are called "perihelion" and "aphelion."

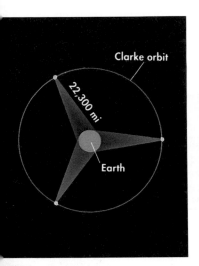

Geostationary, or Clarke, orbit

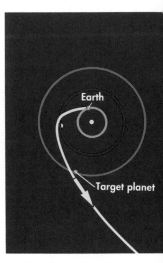

A "slingshot" trajectory

SLINGSHOT ORBITS, which are also called gravity-assist trajectories, are those in which a spacecraft is sent close to a planet in order to use the gravity of the planet to slow down or speed up the craft (usually to speed it up), changing its trajectory. In this way we can send spaceprobes on missions we could not otherwise accomplish because of the limited fuel capacity and power of our space vehicles.

SPACE NAVIGATION

Finding one's way around in space is more complicated than it is on Earth. Since a spacecraft (like an airplane) can move in three sets of directions (forward-backward, right-left, and up-down), there are three numbers needed to determine one's position. On a surface such as the Earth, with no up-down dimension, you need just two, such as latitude and longitude.

RADIO NAVIGATION is commonly used for satellites in orbit around Earth. They are tracked by radio and radar (and sometimes optical telescopes), and thus ground controllers continually know their position.

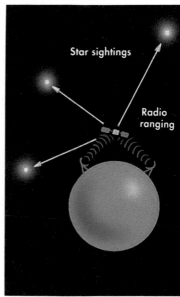

Star sightings

Radio ranging

CELESTIAL NAVIGATION involves observing the relationship of the satellite to stars, planets, and the Sun. Star trackers, which are actually small electronic telescopes, keep known stars in view, and the satellite's computer system calculates its position and orientation from the directions to several stars.

Steering by the stars and by radio

INERTIAL NAVIGATION is a system of devices in the satellite that measures all changes in motion and orientation. It computes where the satellite is based on where it was at some other time.

49

ATTITUDE is the term used to describe the orientation in space of a satellite. Satellites must know their orientation for pointing a camera to take pictures or for pointing a radio antenna back to Earth. Like an airplane, there are three axes around which the craft can turn. Motion around the axis pointing in the direction of flight is called roll. Up-and-down motion, with respect to the orbit, is called pitch. Side-to-side motion is called yaw.

TRACKING NETWORKS of large antennas around the world are used to detect and communicate with spacecraft. Sometimes satellites may be used to locate other satellites, and to relay data from widely spaced Earth stations to a central tracking facility. For NASA, the focus of all tracking activities is at the Goddard Space Flight Center in Maryland.

Principles of inertial navigation

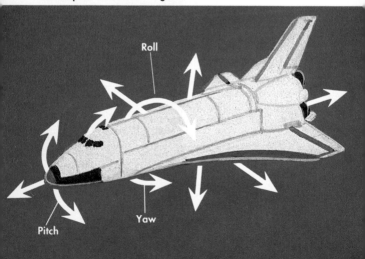

SPACE TRACKING is an around-the-globe, around-the-clock activity. Both the U.S. and the U.S.S.R. military have extensive networks that not only track their own spacecraft but have the additional duty of watching for missile attacks. To do this they must know what is in orbit and where, and they must monitor every launch.

Not much is known (in the non-classified world) about the Soviet tracking system. The U.S. system has its focus inside Cheyenne Mountain, near Colorado Springs, at the joint headquarters of the North American Aerospace Defense Command and the U.S. Space Command. They use a world-wide network of optical and radar tracking equipment that sends in over 45,000 sightings a day for identification and analysis. The optical system is called GEODSS, for Ground-based Electro-Optical Deep-space Surveillance, which uses electronic telescopes located in New Mexico, Hawaii, Korea, Portugal, and Diego Garcia Island in the Indian Ocean. It is said these can detect space objects only a few inches across in low orbits, and only the size of a football in orbits out to about 25,000 miles.

A system of 26 radio and radar sites spaced around the Earth contributes information on objects in orbit. These include the U.S. Navy's Space Surveillance System. In the future, space objects will increasingly be monitored from space by orbiting surveillance satellites.

Private satellite operators—for instance, of commsats— may have their own tracking and control facilities for their satel-lites, and may perform these functions under contract for other satellite owners as well. Such systems are of course much less elaborate than the military systems.

Low-orbit satellites sometimes relay their signals through higher-orbiting satellites. For interplanetary probes, NASA uses its global Deep-Space Network of very large antennas.

ROCKET PROPULSION

Rockets are the technology used for spaceflight. A rocket is a device that carries its own fuel and oxidizer so it can work in a vacuum, unlike a jet engine that uses the oxygen in the air to burn its fuel. A rocket motor works because of Newton's Third Law of Motion (p. 41). The combustion of the fuel and oxidizer produces a force in all directions inside the rocket chamber. One end of the chamber—the nozzle—is open. The burning gases escape out the nozzle, creating an unbalanced force pushing the rocket forward. A toy balloon blown up and released works the same way, with air rushing out one side and the balloon going in the opposite direction. Rockets do not work because their exhaust pushes against the air; in fact, they work better in a vacuum because they do not have air to push out of the way. In some very small rockets, such as those used for minor maneuvering and orientation of a spacecraft, the rocket (often called a thruster) is simply a jet of pressurized (sometimes heated) gas escaping through a nozzle.

Action and reaction make a rocket work

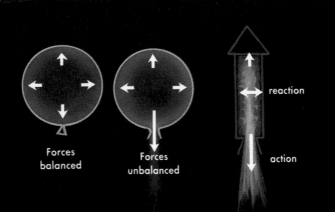

Forces
balanced

Forces
unbalanced

reaction

action

THRUST of a rocket is the amount of push it has, usually measured in tons or pounds of force. For a rocket to be able to take off, the thrust of the motor must be greater than the weight of the rocket. The time during which a rocket is thrusting is sometimes called a "burn."

SPECIFIC IMPULSE measures how efficient a rocket fuel/oxidizer combination is. This is the time during which one pound of fuel can produce one pound of thrust. The most powerful combination in use is liquid hydrogen and liquid oxygen; its specific impulse is around 350 seconds.

SOLID-FUEL ROCKETS use a solid mixture of fuel and oxidizer. The earliest solid fuel was gunpowder, a mixture of sulfur, charcoal, and saltpeter. Today's solid fuels are mixtures of rubbery materials containing powdered aluminum and oxidizer chemicals. They are usually mixed in liquid form and then poured into the rocket casing, where they solidify. The thrust of a solid-fuel rocket depends on the amount burning at one time, which can be controlled by designing the inside of the rocket properly. Once a solid-fuel rocket is ignited, it cannot be shut off; it burns until all fuel is gone.

LIQUID-FUEL ROCKETS use a mixture of liquid fuel and an oxidizer as propellants. Some chemicals must be ignited by a flame or a spark. Others spontaneously ignite when they come in contact; these are called hypergolic propellants. Some common liquid fuels are alcohol, purified kerosene, liquid hydrogen, and hydrazine. Common oxidizers include nitrogen tetroxide and nitric acid, as well as liquid oxygen. A liquid-fuel engine can be controlled in power, shut off, and restarted at will. Supercold propellants like liquid hydrogen and oxygen, called cryogenic fuels, are tricky to handle, requiring the highest technology, but are very powerful.

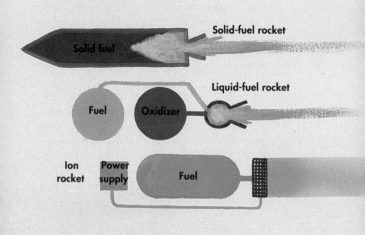

Types of rocket engines

ELECTRIC PROPULSION, sometimes called an ion rocket, is a technique not yet perfected. It works by placing an electrical charge on the molecules of a single fuel, which can be water, hydrogen, mercury, or many other chemicals. There is no combustion, so no oxidizer is needed. The charged molecules (ions) are propelled out of the rocket by a strong electrical field, producing thrust. The total thrust produced by ion rockets is very low, but they use fuel with great efficiency and can function for a long time. In the future they may be used for interplanetary and maybe even interstellar probes.

MASS RATIO is the ratio of the weight of the payload (what you want to get into space) to the total fueled weight of the rocket. This is often very small, only a few percent. For this reason, rockets are usually built to operate in stages.

STAGING is the technique of building a rocket in sections, or stages, each of which provides thrust for a time and then is discarded. The reason for this is to save weight, for the fuel tanks and rocket engines themselves are very heavy. Typical rockets (often called launchers) have two to four stages. The first stage usually burns for only a couple of minutes, getting the rocket going and lifting it up several miles through the thickest layers of the atmosphere. After its fuel tanks are empty, it drops away and a second, smaller stage ignites, carrying the rocket higher and faster. Third and fourth stages may be used to continue the process until the payload reaches the proper altitude and speed for its mission. Stages after the first one are called upper stages.

EXPENDABLE LAUNCH VEHICLES (ELVs) are launchers, or rockets, that are used once and their parts discarded. Most launchers fall into this category. Most are launched from launch pads; a few small rockets are launched from aircraft. So far, the U.S. and Soviet space shuttles are the only reuseable spacecraft.

TO ACHIEVE ORBIT the satellite must get both high enough and fast enough to stay there. For this reason there is often a period of coasting between the burning of the rocket's stages. To place a communications satellite in a geosynchronous orbit, a rocket will typically use three stages to first place it in a highly elliptical transfer orbit with a perigee of 200 miles and an apogee at the geosynchronous altitude of 22,300 miles. The third stage may be called a perigee kick motor or a Payload Assist Module (p. 73). After the satellite has been checked out following several of these orbits, and after it has reached its apogee, a solid-fuel rocket fires to circularize the orbit at the proper altitude. This rocket in the base of the satellite is called an apogee kick motor.

G FORCE is a measure of the acceleration produced by a rocket. One g, or "gee," is the force of gravity. The space shuttle crews and cargo experience a peak force of about 3 g's. An unmanned rocket may reach 11 g's or more.

COST OF LAUNCHING is still very high. In the 1980s, it cost roughly $25 to $150 million to place a few-thousand-pound expendable-rocket satellite into Earth orbit. This translates into costs of $10,000 to $50,000 per pound. And this covers only the costs of building the rocket, the fuel, the payload preparation, and so on. It doesn't include the cost of the payload itself, or such developmental costs as designing the rocket or building the launch pad. One of the major goals of all space programs is to reduce the cost-per-pound to orbit.

Several stages are used to reach orbit

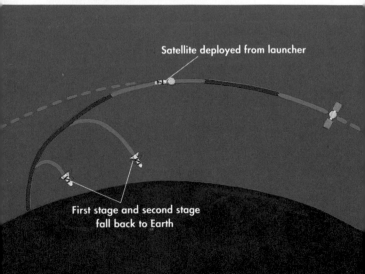

Satellite deployed from launcher

First stage and second stage
fall back to Earth

U.S. LAUNCHERS

At the end of World War II, both the U.S. and the U.S.S.R. captured German rockets, rocket parts, and rocket scientists and engineers. These then became the basis of both nations' rocket programs.

THE V-2 ROCKET, also called the A-4, invented in the early 1940s by a team of German engineers working under Dr. Wernher von Braun, was the world's first rocket to reach space. It was 46 feet high and 66 inches in diameter, weighed 27,000 pounds, and used alcohol and liquid-oxygen fuel to generate a thrust of 56,000 pounds, carrying a one-ton payload 200 miles downrange and 60 miles high. During the last years of World War II it was used to bombard England. Later, U.S. scientists used several to set new altitude records.

V-2 rocket   ORDWAY

JUPITER-C, or Juno I, put the first American satellite, Explorer 1, into orbit January 31, 1958, and carried the first two U.S. manned suborbital spaceflights. Juno I was 71 feet high and almost 6 feet in diameter. It had a liquid-fueled, 83,000-pounds-of-thrust first stage, and three solid-fueled upper stages.

VANGUARD was the second U.S. rocket to orbit a payload. It stood 72 feet high, 3.7 feet in diameter. The first stage used kerosene and liquid oxygen to produce 28,000 pounds of thrust, topped by a second stage using fuming nitric acid and dimethyl hydrazine propellants and a solid-fueled third stage.

SATURN ROCKETS were developed for the manned space program. The Saturn Ib, with 1.3 million pounds of thrust, placed the Apollo-7 crew into Earth orbit, and later launched three Skylab crews.

The three-stage Saturn V, the largest rocket ever made by the U.S., was 363 feet high, 33 feet in diameter, and produced 6.4 million pounds of thrust. It was used for the Apollo program of lunar flights, and a converted third stage became the Skylab space station.

A Jupiter-C rocket   NASA

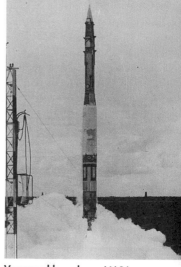

Vanguard launcher   NASA

Saturn V, the Apollo rocket
NASA

ROCKET LAUNCHERS

Scout Ariane-4 H-2 Long March Delta Atlas-Centaur Saturn V

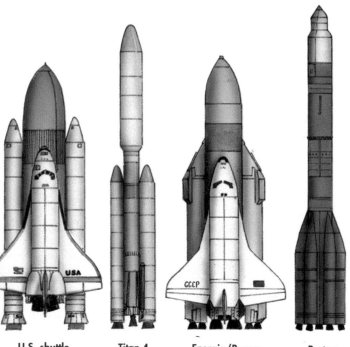

U.S. shuttle   Titan 4   Energia/Buran   Proton

SCOUT is the smallest U.S. satellite launch vehicle. It is also used as a sounding rocket—those that take small payloads above the atmosphere and back again. It was the first American all-solid-propellant launcher. There are many versions of the Scout, but a typical configuration has four stages and stands 75 feet high, 3.7 feet in diameter, and weighs 47,000 pounds. The first stage has 109,000 pounds of thrust; the second stage provides 64,000 pounds of thrust; the third stage is an 18,700-pound-thrust motor; and the fourth stage produces 5,700 pounds of thrust. In this arrangement, Scout can place a payload of about 425 pounds into an orbit 300 miles high.

Its first successful launch was of Explorer 9 in 1961. Scouts have carried such satellites as the Small Astronomy Satellite , the Meteoroid Technology Satellite, and several amateur radio satellites. Scouts have been launched from all U.S. launch sites, most often Wallops Island, and from many other nations.

Scout launcher   NASA

Atlas launcher   NASA

ATLAS was the first operational U.S. ICBM (Intercontinental Ballistic Missile), beginning in 1959. These missiles were gradually withdrawn from service and converted to satellite launchers.

Atlas is called a "stage-and-a-half" rocket because it uses a single main engine together with two booster engines; all are powered by liquid fuel. The three engines are ignited on lift-off, and the boosters fall away halfway through the first-stage burn.

An Atlas-D launched America's first manned orbital flight with John Glenn in 1962. Others have carried such satellites as Mariner, Ranger, Surveyor, High Energy Astronomical Observatory, and many military payloads.

Several versions of Atlas carry either the Agena or Centaur upper stage. The Atlas-Centaur configuration is the current commercial launch vehicle. An Atlas-Centaur is 132 feet high, 10 feet in diameter, and can place 13,000 pounds into low orbit, 4,900 pounds into a transfer orbit, or send 2,600 pounds into an interplanetary orbit.

Mercury-Atlas   NASA

DELTA launchers come in several configurations, differing mainly in the number of solid-fuel strap-on booster rockets, and in what upper stage is used. The Delta 3920 uses nine strap-on motors totaling 766,000 pounds of thrust on the liquid-fueled first stage, which itself produces 205,000 pounds of thrust. The liquid-fueled second stage provides 9,800 pounds of thrust. The third stage produces 18,500 pounds. The payload itself may contain a Payload Assist Module (p. 73).

Overall, the rocket stands 116 feet high, is 8 feet in diameter, and when fueled weighs 426,000 pounds. It can place 2,750 pounds of payload into geostationary transfer orbit and more into low Earth orbit.

An improved Delta model is now in use as the Air Force's Medium Launch Vehicle and for commercial launches.

A Delta commercial launcher MDSSC

TITAN rockets, which began flying in 1959 as ICBMs, are currently the most powerful unmanned U.S. launchers. The Titan Model 3B has two main stages and an upper stage that depends on the payload. Titan-3C and -3D add two solid-fuel boosters to the first stage. Titan-3E is a NASA version used to hurl the Viking and the Voyager interplanetary probes into escape orbits.

The Titan 4, also known as the Titan 34D-7, is 204 feet long, with a payload 40 feet long and 16 feet in diameter. It is capable of lifting 40,000 pounds to low orbit or 10,000 pounds to transfer orbit. Its two first-stage solid boosters have 1.6 million pounds of thrust each, added to 546,000 pounds for the liquid-fueled engine. Second stage provides 104,000 pounds of thrust, and a Centaur upper stage adds 33,000 pounds of thrust. Titan can also carry instead an Inertial Upper Stage (p. 75).

The Titan launcher   NASA

Space shuttle on its crawler transporter   NASA

SPACE SHUTTLE, officially called the Space Transportation System, is the world's first largely reusable spacecraft. There are three major components.

THE ORBITER vehicle looks like an airplane and is about the size of a Boeing 737 jet, with a length of 184 feet and a wingspan of 78 feet. A two-level crew compartment in the nose holds up to seven crew members. Amidship there is a 60-foot-long, 15-foot-wide cargo bay (big enough to hold a bus, or four medium-sized communications satellites) that can carry up to 65,000 pounds of payload into a low-inclination orbit or 32,000 pounds into a polar orbit. Typical flight altitudes are 200 miles up, with a maximum of about 600 miles. An airlock exits into the payload bay so astronauts can perform tasks in space. A Canadian-built Remote Manipulator Arm is used to pick up and retrieve satellites from the cargo bay. The orbiter is powered on launch by three liquid-fueled motors, providing 375,000 pounds of thrust each, as well as by smaller rockets for maneuvering in space.

SOLID ROCKET BOOSTERS (SRBs), each providing 2.65 million pounds of thrust, are ignited at lift-off along with the space shuttle's main engines on the orbiter. They burn for a little over two minutes, by which time the shuttle is about 28 miles up. Then they detach and parachute into the ocean, to be recovered by ships, towed to shore, refurbished, and used again. Each booster is 150 feet long, holds 1.1 million pounds of fuel, and is designed to be used about 20 times.

Inside the space shuttle

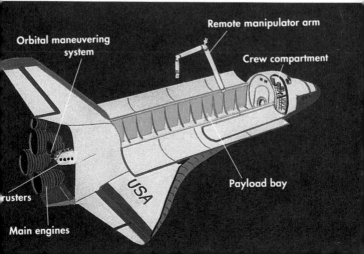

Remote manipulator arm

Orbital maneuvering system

Crew compartment

rusters

Payload bay

Main engines

USA

A space shuttle heads into space
NASA

THE EXTERNAL TANK holds the liquid hydrogen and oxygen fuel for the orbiter's engines. Both the orbiter and the solid rocket boosters are attached to it, and it is the only major part not reused. After 8.5 minutes of flight, the tank is empty and is jettisoned to fall back into the ocean and sink. The tank is made of aluminum; it is 155 feet long, 28 feet in diameter, and 1.65 million pounds when fueled.

A shuttle is assembled in the Vehicle Assembly Building several miles from Launch Pads 39A and 39B at the Kennedy Space Center, or at Vandenberg Air Force Base. The orbiter and the external tank are attached to the boosters, which stand on a Mobile Launch Platform. The whole "stacked" shuttle is then moved by a giant tractor to the launch pad.

At 6.6 seconds before liftoff the main engines ignite and build up to full thrust, which signals the boosters to ignite, and the shuttle takes off. The boosters drop away two minutes later, and the shuttle continues to climb using its main engines. The tank separates at eight minutes into the flight, falling into the ocean, while the orbiter continues on into orbit using its small Orbital Maneuver-

Space shuttle lands like an airplane   NASA

ing System rockets. When its mission is complete, the shuttle is turned around: the maneuvering engines are fired to slow it down, then the orbiter flips over to point nose first.

The shuttle then reenters the atmosphere, getting very hot as it reduces speed using the drag of the air. It is completely unpowered, but has the ability to fly several hundred miles to either side of its orbital path. It lands like a glider on a runway.

After landing, unused fuels are removed and the orbiter is refurbished for another flight. If the shuttle must be transferred some other place for launch, it is carried on the back of a modified 747 airplane.

The destruction of the space shuttle Challenger and death of its crew on January 28, 1986, led to a several-years halt in the shuttle program and left only three operational orbiters, named Columbia, Discovery, and Atlantis. A new one, Endeavour, has now joined them.

A space shuttle seen by a nearby satellite   NASA

THE NATIONAL AEROSPACE PLANE now under study is planned as a hypersonic winged craft that could take off like an airplane from a runway, accelerate to orbital velocity, maneuver in low orbit, and then return to Earth, landing like an airplane again. Such a vehicle could provide access to low orbits at a cost per pound of payload much lower than ordinary rockets or the space shuttle. A commercial version of it may be used as a transoceanic airplane, nicknamed the "Orient Express," able to travel from San Francisco to Tokyo in about three hours.

Requiring very high-strength, lightweight, heat-resistant materials, several designs are being studied for construction in the late 1990s. One calls for using methane as a fuel, with atmospheric oxygen while in the lower atmosphere, then switching to liquid hydrogen and liquid oxygen at higher altitudes. Other plans consider the use of ramjets and supersonic combustion ramjets (called scramjets).

Conception for a U.S. aerospace plane

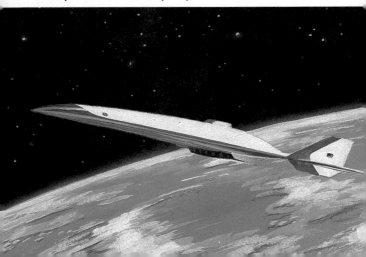

AGENA is the oldest U.S. upper stage, used since 1959 more than 300 times with great success. It has been used as the upper stage for Atlas, Thor, and Titan 3B launchers.

Several military spy-satellite payloads are based on Agena. An Agena was the first space-docking target, and other Agenas launched such payloads as the Mariner-Mars and Mariner-Venus probes.

The latest version, the Agena-D, can be used with a wide variety of payloads, and its liquid-fueled engines can be stopped and restarted in space for versatile maneuvering.

Some Agena stages have been used for missions lasting more than six months in orbit. This stage is 23 feet long, 5 feet in diameter, and weighs 15,000 pounds, 91 percent of which is fuel.

Atlas rocket with Agena upper stage   NASA

CENTAUR upper stage has been used with Atlas and Titan launchers to orbit such Earth-orbit payloads as large communications satellites and orbiting astronomical observatories. It has also been used for many interplanetary missions, including Pioneer, Viking, Mariner, and Voyager.

A typical Centaur stage is 30 feet long and 10 feet in diameter; it burns 31,000 pounds of cryogenic fuel in its two engines, producing 33,000 pounds of thrust. These engines can be restarted in space for maneuvering. Centaur, when used as the upper stage with an Atlas booster, is capable of putting 13,000-pound payloads into low orbits, 4,900 pounds into a geostationary transfer orbit, and, with the addition of a small "kick motor" on the payload, 2,600 pounds into an escape trajectory.

The Centaur contains a very capable guidance, navigation, and systems-control computer that governs not only the Centaur itself, but also the Atlas booster.

Atlas with a Centaur upper stage
NASA

PAYLOAD ASSIST MODULE (PAM) is designed as a small separate upper stage attached to satellite payloads on Delta and Atlas rockets, and for boosting payloads from the low orbit of the space shuttle to geostationary transfer orbit.

When used with the shuttle it is sometimes called the Spinning Solid Upper Stage because directional stabilization is achieved by spinning the PAM and its payload about 50 revolutions per minute.

When used in the shuttle, the spinning is started by electric motors on the structure holding the PAM and its payload. They are then ejected from the shuttle's payload bay by springs, and then allowed to drift many miles away from the shuttle (for safety) before the PAM is ignited to carry the payload higher into space.

When used with expendable rockets, small thruster jets around the PAM cause the spinning.

The first PAM was used in 1980 with a Delta launcher.

PAMs come in different versions for use with shuttle and expendable rockets. PAMs for shuttle use have slightly shorter rocket nozzles so they will fit within the cargo bay.

The PAM is capable of taking almost 3,500 pounds of payload from low orbit to a Clarke orbit.

Payload Assist Module attached to a satellite   NASA

TRANSFER ORBIT STAGE was developed with private funds to provide an upper stage between the capabilities of the PAM, the Centaur, and the Inertial Upper Stage. Weighing 24,000 pounds when fully fueled, it is 11 feet long and 11 feet in diameter, and can take payloads in the range of 6,000 to 13,000 pounds from shuttle orbit to geostationary transfer orbit. Its rocket engine provides 60,750 pounds of thrust. It can also be used as the upper stage for a Titan, taking a payload capacity of 9,000 pounds to Clarke orbit.

APOGEE AND MANEUVERING STAGE is an upper stage 10 feet in diameter and 5.6 feet in length, weighing 9,400 pounds. When used by itself as an upper stage from the shuttle, it can boost 5,300 pounds from low orbit to geostationary orbit. It will also be used for ferrying payloads from the shuttle to the space station. Combined with the Transfer Orbit Stage, the payload capacity to high orbit is increased to 18,600 pounds.

Transfer Orbit Stage   OSC

INERTIAL UPPER STAGE was designed by NASA and the Air Force to launch heavy payloads from both the shuttle and the Titan. It was first used with a Titan in 1982.

It is a two-stage solid-fueled vehicle, 16.5 feet long and 9.6 feet in diameter. Its first stage provides 42,600 pounds of thrust and is connected to the second stage by a structure called the interstage. The second stage gives 17,430 pounds of thrust. The payload, attached to this, contains a sophisticated guidance and control computer. Total weight (without payload) is 32,000 pounds. It can carry a 5,000-pound payload from low orbit to Clarke orbit.

The Inertial Upper Stage   NASA

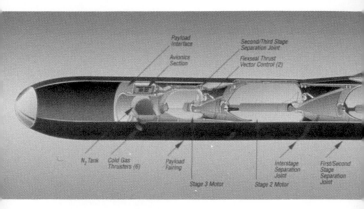

Pegasus air-launched rocket   OSC

PEGASUS is the newest and smallest of United States launchers. It is unique in several ways. First, it is the only launcher developed purely by private industry, albeit using many parts from previous spacecraft. Secondly, it is launched not vertically from a launch pad, but horizontally from under the wing of an airplane. Pegasus is designed to carry small payloads—sometimes called smallsats, or lightsats, or cheapsats—of less than about 950 pounds inexpensively into low orbits a few hundred miles up.

The three-stage, solid-fueled, 50-foot-long Pegasus rocket is attached under the wing of a B-52 bomber and flown to an altitude of about 40,000 feet over the ocean. Pegasus is then dropped from the airplane and its first-stage motors ignite, providing 109,000 pounds of thrust. When the first stage is empty, it drops away and a second stage provides 27,600 pounds of thrust for several minutes before it, too, drops away. A third stage, with 9,800 pounds of thrust, puts the satellite into low Earth orbit.

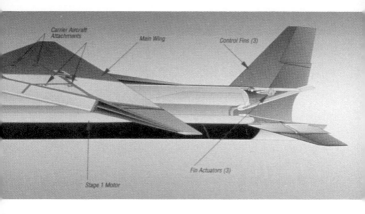

Carrier Aircraft Attachments — Main Wing — Control Fins (3) — Fin Actuators (3) — Stage 1 Motor

SOVIET LAUNCHERS

Like the U.S., the Soviet rocket program following World War II got its start using captured German scientists and rockets. Unlike the United States program, however, the Soviet program was carried on in great secrecy, some of which still surrounds it today. For example, photographs of some Soviet rockets were not made public until 20 years after they were used, and, until recently, some launch sites officially did not exist.

Soviet boosters have tended to be capable of putting heavy payloads into orbit. One reason for this seems to be that Soviet scientists were not able, in the early days of ICBMs, to reduce the weight of their bombs and so had to use more powerful rockets. It also requires more powerful rockets to place satellites into low-inclination orbits from the Soviet high-latitude launch sites. A continuing reason may be that Soviet microminiaturization technology is less advanced, meaning their satellites are heavier. At the present time, they have the world's most powerful booster, Energia, as part of their fleet.

There are three different nomenclatures for Soviet rockets. The Library of Congress system uses letters of the alphabet roughly in order of appearance. Another system, used by the U.S. Department of Defense, uses the prefix SL, meaning "satellite launcher," and a number. Then there is what the Soviets call them.

SOVIET A SERIES launchers, originally ICBMs, launched the first three Sputniks, and include the A1 (or SL-3, or Vostok), now being replaced with more advanced models. Used for heavy spy satellites, they stand 125 feet tall and can place six tons in orbit using a main-core rocket and four boosters.

A2 (SL-4), called Soyuz by the Soviets, has been extensively used for manned launches of Soyuz spacecraft and Progress supply vehi-

The Soviet A1 rocket, Vostok

cles. It stands 163 feet high and can place 7.5 tons into low orbits.

A2e (SL-6) adds an additional upper stage to propel payloads to escape orbit. It has been used for interplanetary probes, but is now used mostly for commsats and deep-space probes.

THE B1 LAUNCHER, or SL-7, is no longer used. It was used to place small payloads in the Kosmos and Interkosmos series into orbit. These satellites had maximum weights of about 1,300 pounds.

THE C1 (SL-8) booster is the only one of the many Soviet launchers known to be launched from all three Soviet launch sites. The first stage develops 176 tons of thrust. Overall length is about 105 feet. They are primarily used now for navsats (navigation satellites) and commsats (communications satellites).

Soyuz   TASS

The Soviet D1 launcher

D SERIES LAUNCHERS have six first-stage engines producing a million pounds of thrust. D1 (SL-13) adds an additional upper stage. This workhorse booster launched the heavy Salyut and Mir space stations, as well as many heavy Kosmos satellites. It can place 20 tons in low orbit.

PROTON, also called D1e and SL-12, is the D1 with an added upper stage to allow escape from Earth orbit. It can carry 5.7 tons as far as the Moon, 5.3 tons to Venus, 4.6 tons to Mars, and 2 tons to Clarke orbit. The Soviet Union has tried to make this launcher available as a commercial vehicle.

There are no Series E, G, or H Soviet launchers.

F SERIES LAUNCHERS include the F1 (SL-11), which with an upper stage can place up to about 9,000 pounds in low orbit. It stands

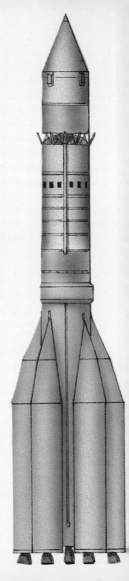

Proton NOVOSTI

149 feet high, is 10 feet in diameter, and produces a thrust of 494,000 pounds in the first stage, 220,000 pounds in the second stage. This launcher has been used almost exclusively for military payloads.

THE F2 (SL-14) adds still another upper stage, so that this launcher can take 12,000 pounds to low Earth orbit. It is about the same size as the F1, but with slightly improved thrust. Typical payloads include the Meteor series weather satellites and probably heavy electronic spysats (spy satellites).

THE MEDIUM-LIFT VEHICLE, also known as the J1, the SL-16, and the SL-X, is a newer Soviet launcher that fills the payload-capacity weight-gap between the capabilities of the A2 and the D1. Its first launch was in 1985. It can boost 15 to 20 tons into low orbit.

ENERGIA, or SL-W, first launched in 1987, is the world's most powerful launch vehicle. It is capable of putting about 220,000 pounds into low Earth orbit.

The launcher stands 198 feet tall, weighs 4.4 million pounds, and has 6.6 million pounds of thrust from its first stage. This consists of a core tank, much like the external tank of the space shuttle, with four liquid-hydrogen/liquid-oxygen engines. Around the core are four boosters using kerosene and liquid oxygen.

There are versions of the Energia for use with unmanned payloads and for the Soviet space shuttle and spaceplane. The payload is carried piggyback on the side of the core vehicle. With the addition of upper stages Energia may be used to build a large Soviet space station in the 1990s.

The booster rockets are reuseable. They separate from the core stage when the launcher reaches about five times the speed of sound. They parachute to Earth and are refurbished for use.

THE SOVIET SPACE SHUTTLE (or SL-W Shuttle, since it is launched on the Energia booster) is very similar to the U.S. shuttle. The Soviet Union made much use of American technology and experience. The flight of the first Soviet shuttle, named Buran ("Blizzard"), was in November 1988.

There are several differences between the Soviet shuttle and the American one. The Soviet shuttle carries its main-stage engines on the external tank, giving improved performance to the orbiter because of reduced weight. The boosters are liquid-fueled. The Soviet shuttle can operate either with a crew or as an unmanned vehicle. For transport between launch sites both the orbiter and tank can be carried piggyback on an airplane, similar to the way the American orbiter is carried on a Boeing 747.

THE SOVIET SPACEPLANE is a smaller vehicle, designed probably for a three-man crew, and capable of rapid launch. It will be used for replacing Mir space-station crews, servicing satellites in low orbit, and for manned reconnaissance missions. The U.S. Department of Defense calls it a "space fighter."

The Soviet Energia and Buran space shuttle

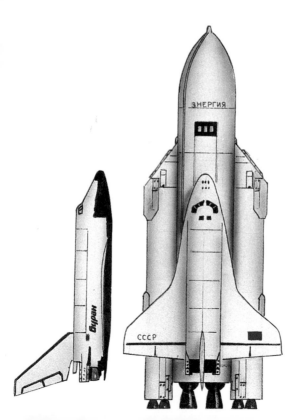

EUROPEAN LAUNCHERS

THE EUROPEAN SPACE AGENCY (ESA) is a consortium of more than a dozen nations that pool their money and expertise on space projects. The biggest contributors to ESA are the United Kingdom, France, Germany, and Italy. Having no launch sites for large rockets in Europe, ESA uses the French Guiana Space Center.

The Ariane, ESA's launcher, had its first flight in 1979 and became operational in 1982. Since then it has captured about half of the world's commercial launch market.

ARIANE-3, one of the versions recently used, stands 162 feet high; it uses four liquid-fueled engines plus two solid-fueled strap-on boosters. The second stage has a single engine similar to those in the first stage, and the third stage uses cryogenic fuels. Total launch thrust is 850,000 pounds, and the vehicle weighs 530,000 pounds. It can place 5,700 pounds into a geostationary transfer orbit, or 3,800 into an escape orbit.

ARIANE-4 is the latest version. The first stage is similar to the Ariane-3, but it is 23 feet longer in order to hold more fuel. By itself it has a thrust from its four engines of 601,000 pounds. The second and third stages are strengthened versions of those from the Ariane-3.

Ariane-4 comes in several configurations. Model 42P, with two solid-fuel boosters, can place 2.6 tons into Clarke orbit. Model 44P uses four solid boosters for a payload capacity of 3 tons. Model 42L uses two new liquid-fueled boosters for 3.2 tons of payload. Model 44LP uses two solid- and two liquid-fueled strap-ons for a 3.7-ton capacity. Model 44L, using four liquid-fueled boosters with a lift-off thrust of 1.2 million pounds, can send 4.2 tons to geostationary orbit.

Ariane-3 launcher   ESA

Ariane-44LP launcher   ESA

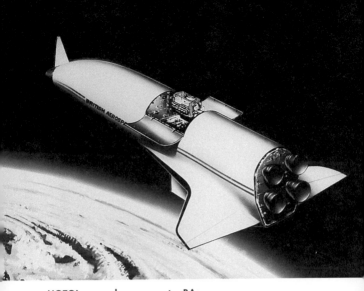

HOTOL spaceplane concept   BAe

HOTOL, for HOrizontal Take-Off and Landing, is a British proposal for a single-stage-to-orbit, 200-ton unpiloted spaceplane that will take off like an airplane from a runway on a rocket-assisted trolley. It will burn liquid hydrogen from on-board tanks with atmospheric oxygen during flight in the lower atmosphere. When the air becomes too thin at higher altitudes, it will switch to on-board tanks of liquid oxygen. Its cargo bay will be about the size of that of the U.S. space shuttle, and the cargo load around 10 tons. When used to carry people, a crew compartment will be fitted into the cargo bay.

After delivering its payload to low orbit it will reenter the atmosphere and land like an airplane. It could be modified to be a hypersonic commercial airliner. HOTOL may be flying by the end of the 1990s.

HERMES is the European Space Agency's project for a reuseable, small space-shuttle vehicle. Present plans call for a three-person crew. Hermes will carry 3 tons of payload. Its overall weight will be around 21 tons.

ARIANE-5, now in the planning stages, will become the launch vehicle for Hermes in the late 1990s. It will also carry unmanned satellite payloads, capable of carrying 15 tons to low Earth orbit and 8 tons to geosynchronous orbit.

A single cryogenic first-stage engine will provide 80 tons of thrust. Two solid-fuel strap-on boosters will each add another 500 tons of thrust. At lift-off the launcher will weigh about 145 tons. For small- to medium-sized payloads there will be a second stage. For missions that require more power, a third stage developed from the Ariane-3 will be used. For manned missions, the Hermes space shuttle is to be the upper stage.

Hermes spaceplane on Ariane-5   ESA

Sänger and Horus concept   ESA

SÄNGER/HORUS is a proposed German two-stage space-plane that is to be fully reuseable. The first-stage Sänger vehicle itself will be a hypersonic aircraft powered by turbo ramjets with an 83-foot wingspan, a length of 165 feet, and a takeoff weight of about 350 tons. It will carry the Horus orbital vehicle piggyback, taking it to an altitude of about 20 miles and a speed of 7,500 miles per hour. Sänger will then separate from the orbital vehicle and fly back to land at an airport for reuse.

Horus, weighing 50 tons, will then continue into space. Payload capacity to low orbit is planned to be either 9,000 pounds of cargo or 4,500 pounds of cargo and a crew of ten. Its mission over, Horus would reenter the atmosphere and land like an airplane.

JAPANESE LAUNCHERS

Japan became the fourth space nation in 1970.

LAMBDA SERIES rockets are small, unguided solid-fueled launch vehicles with a first-stage thrust of 81,000 pounds.

MU SERIES LAUNCHERS include the three-stage Mu-3S, which has two strap-on solid-fuel boosters, has a take-off thrust of 431,000 pounds, and can place 1,700 pounds in low orbit or 300 pounds in an escape orbit. It launched Japan's first deep-space mission to Comet Halley.

N-SERIES LAUNCHERS use a licensed derivative of the U.S. Delta rocket as a first stage. The second stage uses liquid fuel, the third stage is solid-fueled. The N-2 has a total lift-off power of 638,000 pounds. It can lift 770 pounds to Clarke orbit.

The Japanese Mu launcher

THE H-1 LAUNCHER, like the N-series, uses a Thor-derived first stage with nine strap-on solid-fuel boosters. Total lift-off thrust is 640,000 pounds. The second stage uses an all Japanese-design cryogenic-fueled engine producing 22,000 pounds of thrust. Third stage is solid-fueled, with 17,600 pounds of thrust. It stands 132 feet high, and can place 7,100 pounds into low Earth orbit, or 1,200 pounds into geostationary orbit.

THE H-2 LAUNCHER, designed for use after 1992, will use a cryogenic first stage with two strap-on boosters for a total launch thrust of about 130 tons. The second stage will be the same stage as in the H-1. It will be capable of placing 4,400 pounds into Clarke orbit, and will be Japan's most powerful launch vehicle well into the 21st century.

Japanese H-2 launcher

CHINESE LAUNCHERS

The People's Republic of China became the fifth space-capable nation on April 24, 1970. Since then China has launched two to three missions a year. Its first communications satellite in Clarke orbit was launched April 8, 1984. The Chinese have specialized in recoverable unmanned satellites. Some of these have contained experiments performed in weightlessness, and others seem to have been reconnaissance satellites.

LONG MARCH 1, known also as the CZ-1 (Chang Zheng-1) and by the Western nations as CSL-1, was the earliest booster. It was derived from an IRBM (Intermediate Range Ballistic Missile). It has two liquid-fueled stages and a solid-fuel upper stage, together 97 feet tall, capable of putting 660 pounds into low Earth orbit. The CZ-1C is a slightly improved version with a payload capacity of 880 pounds.

Chinese CZ-2C launcher   © Xinhua and New China Pictures

THE FB-1 ROCKET stands for "Feng Bao," which means "Storm Booster." Over a dozen satellites have gone up on this series of rockets. It is a two-stage liquid-fueled rocket that can place 5,500 pounds in a low orbit.

LONG MARCH 2 (CZ-2) is a two-stage rocket, the first stage developing 617,000 pounds of thrust from four liquid-fuel engines. These can put 6,600 pounds into low Earth orbit.

THE LONG MARCH 3 (CZ-3), first flown in 1984, is a three-stage rocket with the same first stage as the CZ-2. It is 142 feet high, weighs 202 tons, and can place 3,100 pounds into a geostationary transfer orbit.

"WEAVER GIRL" (Long March 4) is the newest Chinese launcher, first tested in 1988.

"Weaver Girl"—the Chinese Long March 4 launcher © Xinhua and New China Pictures

INDIAN LAUNCHERS

India entered the Space Age in July 1980. India has used the launch services of both the U.S. and U.S.S.R.

SLV-3, India's first successful launcher, is similar to the U.S. Scout rocket. It placed three satellites into low Earth orbits in the early 1980s.

ASLV, or Augmented Satellite Launch Vehicle, is a 75-foot-high, all-solid-fuel, four-stage rocket capable of placing about 330 pounds into low orbit. It has a lift-off weight of 39 tons and a thrust of 365,000 pounds.

GSLV, Geostationary Satellite Launch Vehicle, is in the planning stages for operation in the mid-1990s. The current design calls for a cryogenic-fueled rocket able to place 2,800 to 3,700 pounds into a geostationary transfer orbit.

India's SLV

ISRAELI LAUNCHER

Israel became the eighth space-capable nation on September 19, 1988, with the launch of its first satellite aboard the Shavit ("Comet") launcher.

SHAVIT is thought to be a derivative of the two-stage solid-fueled Jericho-2 military ballistic missile, which was in turn derived from French rocket research. As a missile it has a range of about 400 miles. As a satellite launcher it placed its first payload into an elliptical retrograde orbit with a perigee of 155 miles and an apogee of 717 miles.

The Israeli launcher   IAI

First Israeli satellite   IAI

The launch was unusual in that it was toward the northwest, over Europe, so that the rocket wouldn't fly over Arab nations. The launch site was the Palmachim Air Force Base.

The 343-pound satellite, named Offeq-1 ("Horizon-1"), is 8 feet long and 4 feet in diameter, and it contains several scientific experiments.

OUT TO LAUNCH

THE LAUNCH PAD, or launch complex, is the location and assemblage of components that holds the rocket ready for launch and supplies the necessary facilities for fueling, electrical power, payload installation, and other support services. Most rockets are attached to the pad by support structures at the base of the first stage. The U.S. space shuttle rests on its solid rocket boosters.

Some rockets, such as the American space shuttle, the European Ariane-4, and many Soviet rockets, are assembled in a separate vehicle-assembly building and then moved to the launch pad several miles away. Most Soviet rockets are carried horizontally on railcars. The American shuttle and the European Ariane-4 are carried vertically. Some other launchers are "stacked" on the pad, and the payload is then installed in the top stage.

Launch pads need to withstand the tremendous heat and vibration of launch. Often huge streams of water are sprayed over the pad during the lift-off. Beneath the pad are flame troughs to carry the rocket flames harmlessly away from the pad.

THE GANTRY is the main structure of the pad, usually higher than the rocket itself. It provides work platforms at many levels, and these can be moved to encase the rocket for work on it and to provide protection from the weather. Similar moving structures, which do not surround the rocket but allow workers and crews access to it, are often called "swing arms."

From the gantry come umbilical lines carrying fuel, air, and electrical power to the rocket and its payload, and carrying back to ground controllers telemetry signals giving the status of all the rocket and payload functions.

COUNTDOWN is the process of preparing the rocket and its payload for launch into space. Fueling of liquid-fueled stages is always done on the pad, whereas the solid-fuel motors are brought to the pad ready for flight. Because the very cold cryogenic fuels boil off, such fuel tanks need to be "topped off" continually until a few minutes before launch. Because fuels are explosive, it is often necessary to vent away any escaping gases. The external tank of the space shuttle has a "beanie cap" that fits over its nose during launch preparation and performs this function.

The launch process is controlled from a launch control center, often called a blockhouse because it is usually underground and built with very thick walls to withstand an exploding rocket. Inside, launch crews monitor the preparation of the rocket and the status of the payload. This information is carried from the pad via wires before launch, and via radio telemetry after launch.

The checkout and the launch of a rocket are usually

Launch pad at Cape Canaveral   NASA

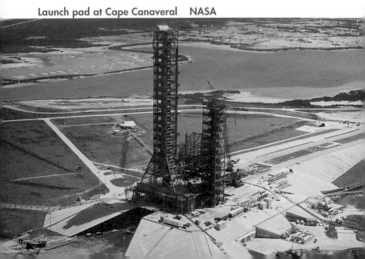

computer-controlled, because there are tens of thousands of measurements that must be made continually. A typical countdown may take about a day. In the last few minutes before launch the safety mechanisms, which prevent the rockets from firing accidentally, are set to allow for ignition, liquid-fuel tanks have their vents closed so pressure that will force the fuels into the engine can build up, and the rocket goes on internal battery power.

At ignition, called "T-0," the rockets are ignited and the cables connecting the rocket with the launch pad are disconnected. Often it takes several seconds for liquid-fueled rocket engines to build up to full thrust. Then solid-fueled boosters are ignited and the rocket begins to "lift off," taking several seconds to "clear the tower."

Several seconds later the rocket turns to the desired direction and angle of flight and continues to climb. As it ascends through the thick lower layers of the atmosphere it is subject to a great deal of stress, so often the engines are reduced in power for a few seconds to avoid over-stressing the rocket. During this stage the rocket is accelerating at several times the force of gravity.

After a minute or two, booster rockets burn out and fall away. When the fuel in the first stage is gone, it drops away, usually falling into the ocean. The second stage ignites and continues to thrust the rocket toward space. Once the rocket is above the thicker parts of the atmosphere, aerodynamic streamlining is no longer needed, so to save weight the nose cone of the rocket, called the payload fairing, is jettisoned.

When the second stage is empty, it separates, and the rocket may coast upward for a while before the third stage fires. The burning of the third stage places the payload into orbit, and it then separates from the rocket. Often small rockets move the third stage aside to prevent its bumping into the payload.

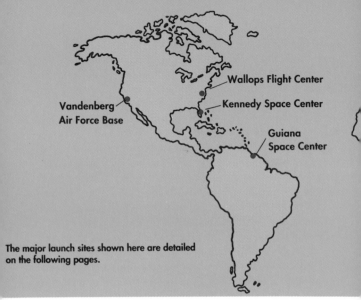

The major launch sites shown here are detailed on the following pages.

SPACEPORTS

Since the most dangerous portion of a launch is the first-stage burn, there must be unpopulated areas in those directions from the launch site toward which rockets are sent. For low-inclination orbits, this means roughly toward the east and southeast; for high-inclination satellites, particularly polar orbits, the direction of launch is toward the north or south.

The latitude of the launch site partially determines how much payload a rocket can carry to low-inclination orbits. A launch eastward from a site close to the equator means a given rocket can carry a greater payload. If two identical rockets were launched from NASA's Kennedy Space Center in Florida (latitude 28 degrees) and from the Guiana Space Center (5 degrees), the latter could carry about 15 percent more payload to Clarke orbit.

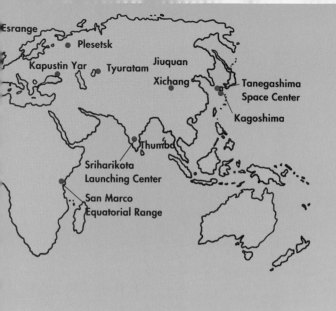

THE WORLD'S MAJOR LAUNCH SITES

	Launch Site	Latitude	Longitude
1	Kennedy Space Center	28.5N	81.0W
2	Wallops Flight Center	37.9N	75.4W
3	Vandenberg Air Force Base	34.7N	120.6W
4	Guiana Space Center	5.2N	52.8E
5	Tyuratam (Baikonur)	45.6N	63.4E
6	Kapustin Yar	48.4N	45.8E
7	Plesetsk	62.0N	40.1E
8	Jiuquan	40.6N	99.8E
9	Xichang	28.1N	102.3E
10	Tanegashima Space Center	30.4N	131.0E
11	Kagoshima (Uchinoura)	31.2N	131.1E
12	Sriharikota Launching Center	13.8N	80.4E
13	Thumba	8.5N	76.9E
14	San Marco Equatorial Range	2.9S	40.3E
15	Esrange	67.8N	20.2E

KENNEDY SPACE CENTER (KSC) is NASA's major spaceport, located on the east coast of Florida on Merritt Island. Next to it is the Cape Canaveral Air Force Station, launch site for the Air Force Eastern Space and Missile Test Range, which extends southeastward across the Atlantic. KSC itself has only two launch pads, 39A and 39B. These were used for all but one of the Apollo launches and for all the shuttle launches so far. All expendable rocket launches and all pre-Apollo manned launches were from the Air Force Station, where most of the pads have now been deactivated. KSC also has a giant vehicle-assembly building and related processing buildings and firing rooms, as well as a 15,000-foot runway for returning space shuttles.

The non-operational areas of KSC are a wildlife preserve. Public tours of KSC are available from the Visitor's Information Center located near the entrance on NASA Causeway. The Center has extensive displays of rockets, satellites, and educational exhibits.

VANDENBERG AIR FORCE BASE (VAFB), also known as the Western Test Range, is located near Lompoc, California. It has been the site of many military missile launches, and of NASA launches of satellites intended to go into high inclination or polar orbits. VAFB has a safe launch direction for polar orbits toward the south, over water. VAFB cannot launch satellites toward the east because first stages would fall in populated areas.

Launch and landing facilities that enable polar orbits for the space shuttle were built here but then deactivated following the Challenger accident in 1986. They will be reactivated sometime in the 1990s.

VAFB is also the site of many operational ICBMs, and of test launches across the Pacific. Hence, this is a heavily classified area not open to the public.

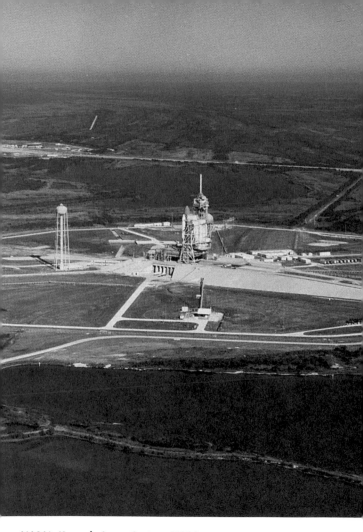

NASA's Kennedy Space Center   NASA

NASA's Wallops Island Facility   NASA

WALLOPS FLIGHT CENTER is located on Wallops Island off the Delmarva Peninsula on the coast of Virginia. It was one of the earliest U.S. launch sites, dating from 1945. Many tests of boosters and other rocket systems occurred there in the early days of the space program. It has safe launch directions to the east and southeast. While many of the flights are of sounding rockets, designed to go up and right back down again, many small satellites have been launched from here using Scout rockets. NASA has launched several non-U.S. satellites from this site.

OTHER NASA CENTERS, which conduct programs and research but do not have launch sites, include the Goddard Space Flight Center, Greenbelt, Maryland; Johnson Space Center, Houston, Texas; Ames Research Center, Mountainview, California; Lewis Research Center, Cleveland, Ohio; Jet Propulsion Laboratory, Pasadena, California; Langley Research Center, Hampton, Virginia; and several that conduct mostly aeronautical research. Some have visitor centers open to the public.

PLESETSK, known also as the Northern Cosmodrome, is located 100 miles south of Archangelsk, U.S.S.R., near the town of Plesetsk. It is the Soviet equivalent of Vandenberg Air Force Base and is the major Soviet military launch site; it is heavily defended by missile emplacements. Plesetsk is by far the world's busiest spaceport, having launched well over 1,000 missions, typically 50 to 70 a year. The site is at least 60 miles long, with dozens of launch pads as well as research facilities. It is ideally positioned for launching polar-orbiting satellites, typically toward the northeast, and it is from here that Soviet reconnaissance satellites are launched.

The existence of this cosmodrome was unknown until a group at a British boys' school tracked a launch in 1966 and calculated that it must have come from this region, not from one of the other space centers then known.

The Soviet manned space control center at Kaliningrad   TASS

TYURATAM (also called Baikonur because the Soviets claimed for many years it was near that town) is really more than 200 miles southwest in Kazakhstan, just east of the Aral Sea. It is near the Sar-Darya River. A newly built city for the base workers, Leninsk, is nearby.

This huge spaceport has more than 80 launch pads, and was the location from which both Sputnik 1 and Yuri Gagarin were launched. Typically, more than a dozen flights a year are launched from Tyuratam, which is the second busiest spaceport in the Soviet Union. All Soviet manned missions, and all interplanetary probes, are launched from here. This is also the site for launches of the Energia booster, and new facilities are being developed for launches of the Soviet shuttle and spaceplane.

KAPUSTIN YAR, also known as the Volgograd Station, is the oldest of the Soviet launch bases; it is located on the Volga River not far from the city of Volgograd. It was here the Soviet and German engineers launched the V-2s captured at the end of the war. This was the first-known of the Soviet rocket bases, early missile launches having been tracked by radar from Turkey to the south.

This spaceport is largely used for military missile tests (particularly antiballistic missile targets), for small scientific satellites in the Cosmos series, and for sounding rockets. Most orbital launches from here have used the B and C series of rockets. More recently it has been the launch site of the small Soviet spaceplane (p. 83). This may signal a coming increase in the usually small launch-rate from Kapustin Yar, depending on whether only the test flights or the operational flights will be made from here.

KALININGRAD is the main control center for Soviet manned space missions.

Cosmonauts at the Tyuratam launch base   TASS

A Soviet training center   TASS

TANEGASHIMA SPACE CENTER, with its Osaki and Takesaki launch sites, is located on Tanegashima Island, about 600 miles southwest of Tokyo, Japan, and 50 miles south of the southern tip of Kyushu, southernmost of the Japanese major islands. The Takesaki site is used for small sounding rockets on research flights. The Osaki launch site is used for orbital launches of larger vehicles. It was first used in 1974 as the launch site of an N-series launcher, and is the site for the new H-2 vehicles.

The island is in the middle of one of Japan's richest fishing areas, and falling first stages of rockets could produce a hazard to fishing boats, requiring the downrange areas to be evacuated for each launch. Hence, an agreement between the Japanese government and the fishing industry restricts launches to February and August. Typically, one or two launches a year are carried out here, but this number may rise as the H-2 launcher comes into operation.

KAGOSHIMA, also called Uchinoura, is on the southern tip of Kyushu Island. It was inaugurated by a first launch in 1964 of a Lambda sounding rocket. (Several earlier Japanese sounding rockets had been launched from the nearby Akita Rocket Range.) Japan's first six satellites were launched from Kagoshima, beginning in 1970. The largest rockets launched from here are the Mu-series. All launches have been of scientific payloads, both sounding rockets and several orbital missions. Typically, there is one launch a year from this base.

OTHER JAPANESE SPACE FACILITIES are at the Tsukuba Space Center north of Tokyo and near Tanegashima; downrange are tracking stations at Okinawa and Ogasawara on Chichi-Jima Island, and a mobile station that can be set up either on Kwajalein Atoll or Christmas Island.

Tanegashima launch site

JIUQUAN, previously referred to as Shuang Chen-Tse, or "East Wind Center," is in northwest China, in Gansu Province, near the border with Mongolia. It was first used in the late-1960s to launch military rockets, some possibly with nuclear warheads, into a test range in the Gobi Desert. From here was launched China's first satellite, China-1, in 1970. Most launches are toward the southeast.

Typical payloads now launched from Jiuquan are Earth resources satellites and those with recoverable payloads. The latter are launched into low orbit for days to weeks, and then reenter under controlled conditions so they may be found and the specimens aboard studied.

Because the Chinese do not have any territories outside their borders for tracking satellites downrange, they use tracking ships sent out into the Pacific.

XICHANG, earlier called Chengdu, is located in China's Szechwan province. The site is more suitable for launches to geosynchronous orbits than Jiuquan is since it has a lower latitude, and all such launches are now made from here.

Although small compared to some of the giant American and Soviet bases, Xichang is being expanded to serve as the launch base for Long March 2 and 3 rockets. China now offers its launch services to other nations and firms needing to place satellites into orbit.

Like the Soviet Union, China has only land-locked launch sites, which restricts the directions in which they can launch their rockets. Despite the apparent availability of seacoast sites, they seem to prefer to keep their launch bases inland for security reasons.

Xichang launch site   © Xinhua and New China Pictures

GUIANA SPACE CENTER, known by its French name Centre Spatial Guyanais (CSG) and by the name of the nearest town, Kourou, is located in French Guiana, on the northeast coast of South America. It was established by France with a first launch in 1968. It is now used by the European Space Agency and its commercial launch organization, Arianespace, to launch Ariane rockets.

ELA-1 rockets are assembled on the launch pad. The newer ELA-2, as well as future ELAs under construction for the Ariane-5, use a vehicle-assembly building, in which the entire vehicle is put together on its launch platform and checked out by a sophisticated computerized system. The platform and rocket are then carried by rail about a mile to the launch site. There are also plans for a landing strip for use with the Hermes spaceplane.

Centre Spatial Guyanais   ESA

India's Shar spaceport

SRIHARIKOTA LAUNCHING CENTER, called Shar, is India's major launch site, located on the east coast north of the city of Madras on the Bay of Bengal. Shar began with sounding rocket launches in 1971, and since then has been the site of most of India's satellite launches. Shar is also responsible for administering several other minor launch sites, including Balasore Rocket Launching Station, located on India's east coast, south of the city of Calcutta. Balasore's more northern latitude makes it less suitable for orbital launches.

THUMBA Equatorial Rocket Launching Station is located just north of the city of Trivandrum near India's southern tip. Thumba is now administered in cooperation with the United Nations and is used for atmospheric sounding-rocket research by many nations, including France, Germany, Japan, and the United States.

SAN MARCO EQUATORIAL RANGE is operated by Italy and the United States. It is located in Formosa Bay, three miles offshore from the coast of Kenya.

The San Marco launch platform is a 3,000-ton, 98- by 382-foot steel structure standing on 20 legs embedded in the seafloor. Scout rockets are assembled and tested in a horizontal position, then raised to a vertical position for launch.

About 4,000 feet from San Marco and connected to it by 23 electrical cables is the Santa Rita control platform, a triangular "Texas Tower"-type platform like those used for oil-drilling rigs. It is 137 feet on a side. A crew of 80 controls and tracks launches from here. Both platforms generate their own electricity.

The most famous satellite launched from San Marco, and the first U.S. satellite launched by another nation, was Explorer 42, better known as "Uhuru," in 1970. During its very successful mission it scanned the sky for X-ray sources, and identified the first suspected black hole.

WHITE SANDS TEST FACILITY is a NASA research laboratory near Las Cruces, New Mexico, next to the U.S. Army's White Sands Missile Test Range. Here NASA tests rocket engines and power systems, particularly for the space shuttle. No launches are made from the NASA facility, but suborbital launches are made from the Army range. These are sounding rockets for both civilian and defense research. The first captured V-2 rockets were launched from here.

WHITE SANDS SPACE HARBOR, a landing strip for use by the returning space shuttle, is also located here. It is for use if the other shuttle landing sites at the Kennedy Space Center and Edward Air Force Base, in California, have bad weather.

Sweden's Esrange launch site

ESRANGE is the world's most northerly launch base, located near Kiruna in northern Sweden, above the Arctic Circle. It is operated by the Swedish Space Corporation and is used by Sweden and many European nations to launch sounding rockets. There are six launch pads available for use; from these sounding rockets can reach altitudes of at least 300 miles. Many of these are designed to study the upper atmosphere, polar phenomena, and the interaction of the Sun and the Earth's magnetic field. About half of the rockets are recovered. Associated with the launch pads are tracking and control facilities, as well as a receiving station for Landsat and Spot photographs. The broadcast satellite for Scandinavian television is also controlled from here.

ANATOMY OF A SATELLITE

Every satellite and spaceprobe has a mission to perform. It will therefore contain various components that enable it to perform its mission. These may include cameras, telescopes, other sensors, radio receivers and transmitters, or other instruments. In addition, whatever the mission, there will also be a number of subsystems common to almost all spacecraft. Because most satellites cannot be repaired once in space, many of the systems are redundant: they have multiple, duplicate components that can be switched in and out as needed in case one fails. Spacecraft components, as well as the entire assembled craft, are subjected to severe tests to ensure reliability once in space. Pioneer 10, for example, launched in 1972 and now outside the solar system, is still sending back data.

THE STRUCTURAL SUBSYSTEM is the body of the satellite. Other components are attached to it. Once in space a satellite has no weight, but it must nonetheless be designed to withstand forces many times the force of gravity during launch and during orbital maneuvers. Consequently, large, flimsy, extended structures, such as antennas or arms holding solar cells, are often not extended to their full length until the craft is on-station or all major propulsive events are over.

POWER SYSTEMS provide electrical energy to the other systems. In almost all satellites the power comes from the Sun, and is converted to electricity by solar cells. These are heavy and may be a major part of the total weight of a spacecraft. It takes about a square yard of solar cells to produce 100 watts of electricity. Solar panels must often rotate to face the Sun, and the bearings that allow this are critical components.

BATTERIES supply power to the satellite during launch, between the time of launch and when the solar cells become useful, and during any periods in which the Sun is eclipsed.

NUCLEAR POWER SUPPLIES have been used in a few special spacecraft, such as Soviet military ocean-observation satellites that need very large amounts of power for their radar systems. Spaceprobes such as Voyager, designed to explore planets far from the Sun, where the sunlight is weak, use radioisotope thermal generators that convert the heat of radioactive materials into electricity.

Voyager spacecraft

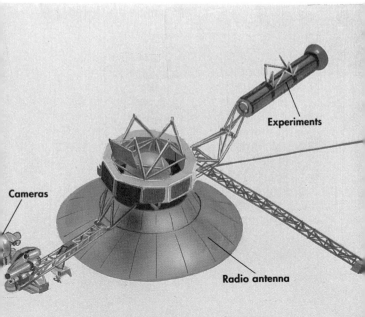

Experiments

Cameras

Radio antenna

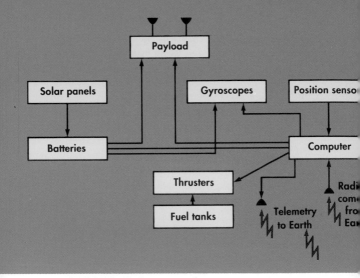

Functional diagram of a spacecraft

THERMAL-CONTROL SYSTEMS keep a spacecraft at the proper temperature. Every square yard of area exposed to the Sun picks up over 1,300 watts of heat, about the same wattage as a steam iron. Spacecraft systems also generate heat. In the vacuum of space there is no air to carry heat away by convection; a craft can only radiate heat away. The spacecraft components must be kept within fairly narrow temperature ranges, usually between about 40 and 100 degrees Fahrenheit.

Passive thermal-control measures include painting the outside a highly reflective white; covering parts with mirrors or gold-plated foil, which is very effective at reflecting solar infrared rays; and painting areas black so they will pick up heat when facing the Sun and radiate heat away when facing away from the Sun.

Active thermal control includes having moveable louvers, white on one side and black on the other, which are turned to control the heat gained or lost. To keep satellite systems warm during eclipses, batteries supply heaters. Some satellites spin continually, a method called "barbecue mode," to keep one side from getting too hot.

THE NAVIGATION SUBSYSTEM is responsible for determining the position and attitude of the spacecraft, and for sending signals to the propulsion system for correction. For Earth-orbiting satellites, position is often determined by ground-tracking, but deep-space probes also use stellar navigation. Objects sighted to determine orientation include the Sun, Earth, planets, and stars. For satellites that must keep antennas pointed toward Earth, such as commsats, an Earth-horizon sensor is often used. Sometimes directional antennas on the satellite track radio transmitters on Earth.

A PROPULSION SUBSYSTEM is required for stationkeeping, keeping the spacecraft in its proper location or trajectory, and for orientation, keeping the craft pointed correctly. Usually these are very small rocket motors that each provide only a few pounds of thrust. They are arranged around the spacecraft in such a way that they can be used either to move the craft in a particular direction or to rotate it about some axis. More powerful systems may be used to change the orbit or trajectory of the craft.

FUEL used for stationkeeping is usually hydrazine. It does not burn as does the fuel for large rockets, but merely escapes under pressure from the rocket nozzles. To provide slightly more thrust, it may be heated in the nozzle. Hydrogen peroxide has also been used.

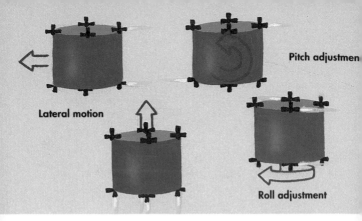

Pitch adjustmen

Lateral motion

Roll adjustment

Satellite thrusters

TRACKING, TELEMETRY, AND CONTROL (TT&C) is the system by which the satellite's operators know where the satellite is and what it is doing, and can command it to do things to fulfill its mission. The telemetry systems monitor the electrical voltages, currents, temperatures, and switch settings within the satellite and report these to control stations on Earth by radio. They may also record data acquired by the satellite for relay to Earth at convenient times. The system uses a special omnidirectional antenna that ensures that, even if the satellite goes out of control and starts tumbling, its operators can always stay in touch with it. When a satellite is onstation and is properly oriented, telemetry signals may also use larger antennas pointed at Earth to allow more data to be sent faster.

TT&C signals are received by a worldwide network of tracking stations. Signals from low-orbit satellites can be received by antenna "dishes" only ten or twenty feet in diameter. Very distant spaceprobes such as Pioneer 10—now close to 3 billion miles from Earth—require dishes a hundred feet in size.

118

RESEARCH SATELLITES have an unobstructed view of Earth, the planets, the Sun, and the cosmos. They include such passive satellites as Lageos, designed to reflect laser beams from Earth in order to determine precisely our planet's shape, and upper-atmosphere research probes that release clouds of gases to stream along the lines of force of Earth's magnetic field. At the other end of the complexity scale are the semiautonomous robotic planetary probes designed to work for years billions of miles from Earth.

"Particles and fields" probes study the interplanetary electrical and magnetic fields, and detect cosmic rays and the solar wind—the stream of high-energy atomic particles thrown off by the Sun.

Orbiting observatories study the cosmos in the visible and invisible wavelengths of the spectrum, most of which never make it through the atmosphere to surface observatories. (Different physical processes produce light of different wavelengths, so each new "window" in the spectrum gives important new information about the universe.)

Planetary probes visit our cosmic neighbors, sending back beautiful photographs as well as data about the environments near them. Solar probes study the star upon which our lives depend.

Earth-orbiting satellites

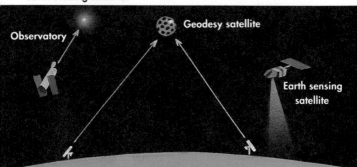

APPLICATIONS SATELLITES make use of the airlessness and weightlessness of space for practical purposes. Remote sensing satellites provide data for better weather forecasts, crop forecasts, the monitoring of forest fires, and other practical uses. Navigation satellites allow ships, aircraft, trucks, and trains to determine their positions very accurately. Communications satellites carry tens of thousands of telephone calls and hundreds of television signals around the globe.

In the future, applied research aboard satellites and space stations will allow the creation of space factories to produce materials of great economic benefit to Earth. Among the things we may develop are new pharmaceuticals for medicine, purer crystals for electronics, and new alloys for building.

The following pages show a few representative spacecraft from a variety of nations.

Spaceprobes have visited almost all the planets

The Hubble Space Telescope   NASA

THE HUBBLE SPACE TELESCOPE (HST) has been called "the most important scientific instrument ever flown." It is named for astronomer Edwin Hubble, who discovered the expansion of the universe.

HST, launched in 1990 by the space shuttle to an orbit about 370 miles up with an inclination of 28.5 degrees, can peer into the universe seven times farther than any previous man-made instrument. It can detect objects 1/50 the brightness detectable from ground-based telescopes. Five sensitive instruments share the light collected by the telescope's 8-foot-diameter main mirror. Despite some initial optical problems, HST is returning much valuable data.

HST weighs more than 20,000 pounds and is 15.5 feet in diameter and 48 feet long. Two large solar panels provide 4,000 watts of electricity. Every few years HST will be visited by the shuttle for refueling, refurbishing, and upgrading.

HST takes electronic pictures and measurements and relays them to the Space Telescope Science Institute located in Baltimore. Here astronomers record the observations and direct its activities.

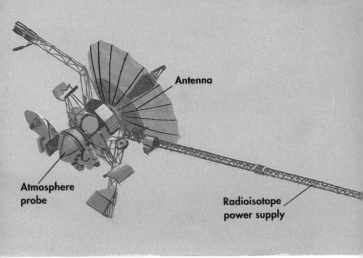

The Galileo Jupiter spaceprobe

GALILEO is now on its way to Jupiter. Galileo's path takes it first into a long elliptical orbit about the Sun, which swings it back months after launch to pass closely by Earth once again to pick up more energy in a slingshot trajectory. It then heads out to Jupiter.

About 100 days before encountering that giant planet in 1995, it will separate into two parts. One goes into orbit around Jupiter, approximately following the orbit of the satellite Ganymede, with a period of about seven days. During 11 orbits it will take close-up photographs of Jupiter and many of its satellites.

The second part of Galileo is an atmospheric probe. It will plunge into Jupiter's atmosphere at a speed of 100,000 miles an hour. A heat shield and then parachutes will slow it down. As it drifts slowly downward, it will radio back a profile of the atmosphere.

The Advanced X-Ray Astrophysics Facility (AXAF)   NASA

AXAF (Advanced X-Ray Astrophysics Facility) will explore the depths of the universe by looking for X rays from cosmic sources. (These wavelengths never make it through the atmosphere to Earth's surface observatories.) It will continue studies begun by the earlier satellites Uhuru and the High Energy Astrophysics Observatories. It will complement the observations of the Hubble Space Telescope.

AXAF will be 14 feet in diameter, 43 feet long, and weigh 10 tons in orbit 300 miles above the Earth. Its main instrument will be an X-ray telescope 4 feet in diameter with 100 times more sensitivity to faint X-ray sources than earlier satellites. It will be able to pinpoint the position of sources four times more accurately in the sky as well. Through international cooperation with NASA, instruments from researchers in the Netherlands and the United Kingdom are being provided. This space observatory is designed to last several years.

123

ULYSSES is a spaceprobe designed to explore a never-before-visited region of our solar system, the huge volume lying above and below the plane of the Earth's orbit (ecliptic) and the polar regions of the Sun. All previous interplanetary missions have been confined to the narrow disk within which all the planets orbit.

The 800-pound spacecraft, provided by the European Space Agency, was launched in 1990. The U.S provided the launch services, the 260-watt power supply, some experiments, and tracking using NASA's large-dish antennas.

The craft is heading toward Jupiter, using a slingshot orbit over the giant planet's pole to swing it into a trajectory that will take it over the poles of the Sun in 1994 and 1995. This mission is designed to last about five years and give us important new information about the three-dimensional structure of interplanetary space around the Sun.

Ulysses spaceprobe studies the Sun's poles

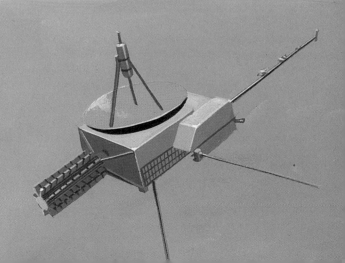

Soviet Phobos spacecraft photographs that satellite

PHOBOS, launched in 1988 by the Soviet Union, was a pair of 10,000-pound spacecraft sent to study Mars and its two satellites, thought to be captured asteroids. They had been first photographed by Mariner 9 in 1971. Cooperative experiments on board came from researchers in France, Austria, West Germany, and Sweden. Phobos 1 failed en route to Mars due to a controller's error.

Phobos 2 entered orbit around Mars early in 1989. The craft was to have studied Mars' inner satellite Phobos, a small world with a very low escape velocity. Lasers from the probe were planned to vaporize a small portion of the surface for chemical analysis. Later the craft was to release two landers, one to drive a probe into the satellite's surface in order to test its properties, another to hop from place to place across the surface for a more complete survey. Unfortunately, the spacecraft failed after sending back only a few photographs and before it made any close studies of the satellite.

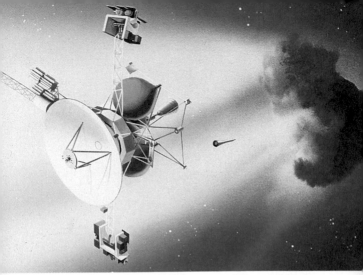

Comet Rendezvous/Asteroid Flyby space mission   NASA

CRAF (Comet Rendezvous/Asteroid Flyby), also called Mariner Tempel-2, will be the first to use a new multipurpose spacecraft structure to which the specific instruments needed for the mission are to be attached.

Plans are to launch CRAF aboard a Titan-Centaur in 1993. It will fly by Venus that August, then swing by Earth again on a slingshot trajectory, picking up speed for its trip outward. In early 1995 it will fly by the asteroid Hestia, giving us our first close look at one of these minor planets.

In 1996 CRAF will approach to within 3,000 miles of Comet Tempel-2. It will then close to within about 600 miles, then down to 20 miles, and fire a penetrator probe into the icy nucleus of the comet to measure its properties. Following the comet as it heads inward toward the Sun and heats up, liberating gases and dust, CRAF will move slowly away from the comet, traveling down the comet's tails.

COMMUNICATIONS SATELLITES (commsats) are the most widely used applications satellites so far, a multi-billion-dollar-a-year business. Almost all commsats—over 100 of them—are in Clarke orbit 22,300 miles above the Earth's equator. At this distance they orbit the Earth in 24 hours, the same time it takes Earth to revolve once. As seen from Earth they appear to be stationary in the sky. Thus antennas on the ground do not need to follow them across the sky.

Each satellite receives radio signals from an Earth station antenna, shifts the frequency, amplifies them, and sends them back to Earth. The circuit aboard the satellite that does this is called a transponder or repeater. Transponders have powers ranging from a few watts to a few hundred watts. Although only about a thousandth of a billionth of a billionth of the power radiated by the satellite makes it to the antennas on Earth's surface, receivers are sensitive enough to allow even small antennas to be useful.

Linking all of Earth via communications satellites

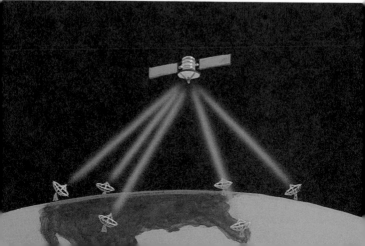

The larger satellites, such as those of the International Telecommunications Satellite Organization (Intelsat), are intended to relay telephone calls, digital data, and television across the oceans. One large satellite can carry 100,000 telephone calls simultaneously, plus several television signals. Other Direct Broadcast Satellites carry television signals directly to small antennas on people's homes. Still others connect ships at sea and aircraft with telephone systems anywhere in the world.

Commsats may have antennas designed to send their signals to the 42 percent of the globe they can see from their orbit, or the antennas may be "spot beams" designed to focus on small areas of the Earth.

Commsats range in cost from a few tens of millions of dollars for the smallest ones to several hundreds of millions of dollars for the largest, most powerful ones. This does not include

An Intelsat VII communications satellite INTELSAT

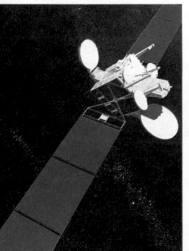

Spacenet commsat in the Clarke orbit GTE

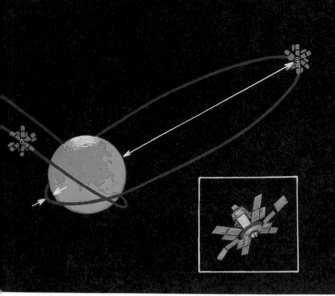

Molniya satellite over northern regions

the additional tens of millions of dollars needed to launch them, or to operate them once in orbit.

There are several commsats not in Clarke orbit. Because it is difficult for a satellite over the equator to communicate with areas at high latitudes, the Soviet Union has a series of commsats, known as Molniya, in long, looping orbits that take 12 hours to go around once. Several of these are spaced around one orbital path, and each is used in turn when it is near its apogee point and hence moving slowly. An Earth station tracks each one for several hours until it begins to move too fast, at which time communications traffic is switched off to another Molniya then nearing its apogee. The U.S. military has a similar satellite system for communicating with its submarines and aircraft when they are near the North Pole.

REMOTE SENSING is the technique of studying something from a distance by analyzing light. Everything emits and reflects light uniquely, depending upon its chemical composition, temperature, pressure, and other properties.

Remote sensing satellites (sometimes called sensats) have a bird's-eye view of Earth and can examine the entire surface every few days to weeks, enabling us to detect changes. Sensor instruments provide data that allows scientists to determine if a crop is under stress, roughly predict crop yields, find the boundaries of a forest fire or a flood, detect minerals, and study the effects of flooding, pollution, or urbanization.

Sea-sensing satellites can determine temperature, wave height, wind direction, and sea level. By knowing the temperature of water preferred by certain fish, satellites can guide fishing fleets to schools of fish quickly.

Weather or meteorological satellites (metsats) provide the photographs you see on television weather reports. They can determine the temperature, winds, weather fronts, cloud pat-

Remote sensing satellites detect radiation from Earth

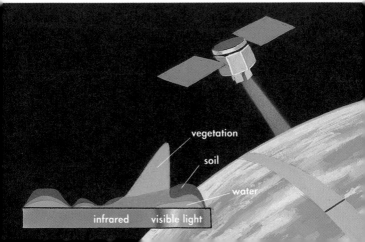

terns, and moisture in the air. They are especially important since most of Earth's surface is ocean, and much of our weather originates there.

RESOLUTION, the ability to detect small objects and tell them apart, is an important characteristics of a sensat. Landsat 5 has a resolution for one of its sensors of 240 feet. This means it can detect areas an acre in size, and tell one acre from another.

The economic benefits of sensing satellites are hard to calculate, but certainly large.

Scanning width

Picture size

Studying Earth from space

Radar "photograph" of Earth from Seasat   NASA

TOPOGRAPHIC RELIEF FROM SEASAT ALTIMETER MEAN SEA SURFACE
JULY 7 - OCTOBER 10, 1978

NW - SE GRADIENT, METERS/DEGREE

-5   0   +5

LANDSAT is the series of U.S. remote land-sensing satellites. The first one was launched in 1972. In the mid-1980s Congress decided to transfer the Landsat program to private industry. The current satellite is Landsat 5, launched in 1984.

Landsat is in a polar orbit at an altitude of 438 miles, oriented so that it looks down and takes photographs of areas 115 miles square. The orbit changes orientation slowly and continually so that it always passes over parts of Earth that have a local time of about 9:30 A.M. This is so that the illumination angle of the Sun is always the same, which allows for comparisons between photographs taken on different days. In 18 days Landsat scans the entire Earth and then begins again.

Landsat weighs almost 4,300 pounds, and is 13 feet long, 6 feet wide, and 12 feet high including its solar panel and antenna. The sensors are (1) a Multispectral Scanner, which detects light in four different wavelength bands with a resolution of 260 feet, and (2) a Thematic Mapper, sensitive to several other wavelengths of light, with a best resolution of 100 feet.

Most Landsat photographs are printed in false colors to enhance visibility. Water shows up as black or blue, urban areas are gray, growing plants are pink or red, desert areas are brown.

A Landsat remote sensing spacecraft   NASA

A Landsat photograph   NASA

SPOT, standing for System Probatoire d'Observation de la Terre, is an Earth resources satellite of the European Space Agency.It was first launched in 1986 by an Ariane rocket, the first polar launch for that rocket.

Spot is roughly cubical in shape, about 6.5 feet on each side. Extending from this main body are solar panels about 50 feet long that provide 1,800 watts of power to the satellite. It weighs 4,000 pounds.

Spot is in an orbit 522 miles up, and repeats its ground track every 26 days. Each image covers an area about 38 miles on a side, with a resolution of 33 feet for black-and-white pictures and 66 feet for color pictures, much better than Landsat. This allows finer detail to be seen on Earth.

Newer, more capable versions of Spot, with higher resolution and more detailed light analyzers, are planned for the l990s.

The Spot remote sensing satellite © 1990 CNES. Provided by SPOT Image Corporation

A photograph of part of Earth from Spot © 1990 CNES. Provided by SPOT Image Corporation

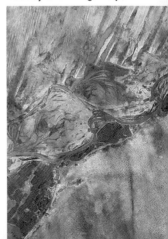

GOES is a series of Geostationary Operational Environmental Satellites. Several GOES are in orbit 22,300 miles up, spaced to give full coverage of the Earth's weather. These satellites also monitor radio reports from thousands of unmanned weather-observing stations and high-altitude meteorological balloons, relaying the data to weather centers on Earth.

The older GOES spacecraft are shaped like hat boxes, about 7 feet in diameter and 11 feet high including their antennas. On orbit they have a weight of 861 pounds. They are spin-stabilized, maintaining their orientation in space like a gyroscope. Their sensors scanned across the Earth's surface as they turned.

The newer GOES satellites are a very different design. They are three-axis stabilized, so they do not spin and their instruments are continually pointed down at Earth. They weigh almost 2,900 pounds. The instruments include new sensors to examine a wider portion of the spectrum; they are more sensitive than those on earlier satellites.

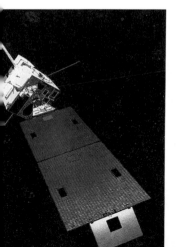

A newer GOES weather satellite NOAA

GOES photograph of Earth's weather   NOAA

A NOAA weather satellite   NOAA

NOAA is a series of low-orbit weather satellites named after the U.S. National Oceanic and Atmospheric Agency, which manages all metsats. They complement the GOES satellites by providing closer, more detailed photographs and measurements of the atmosphere. NOAA satellites are about 500 miles up in polar orbits that take them over all of the Earth. They weigh about 2,000 pounds and are about 13 feet long, not including the solar panels, which supply 1,500 watts.

Their instruments include high-resolution cameras and devices to determine atmospheric temperature. They also relay reports from weather balloons and buoys, and from remote stations.

Aboard the later NOAA satellites are receivers for the Search and Rescue system. These detect distress signals from aircraft that may have crashed in remote areas beyond the range of the usual distress receivers on Earth.

NOAA image of Earth   NOAA

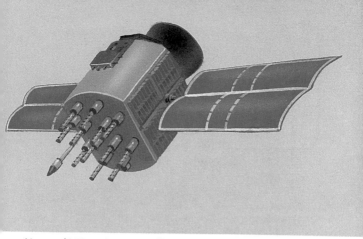

Navstar/GPS navigation satellite

NAVSTAR, or GPS, the Global Positioning System, is a group of navigation satellites designed for use by the U.S. military but also usable by civilian vehicles. There are 18 operational Navstar satellites, six spaced around in each of three 12,500-mile-high orbits with different inclinations, each with a period of 12 hours.

User equipment ranges from small hand-held receivers to larger receivers carried aboard ships and aircraft. These receivers automatically pick up signals from the several nearest Navstars to calculate the receiver's position to within 50 feet in three dimensions, and velocity to within a fraction of a mile per hour.

The current Navstars are about 1,700 pounds in weight, use 700 watts of power from solar cells, and are "hardened" against radiation from nuclear blasts. They also have nuclear detection devices aboard, and can detect if someone has tried to tamper with them either by hitting them or by shining a laser at them.

MANUFACTURING IN MICROGRAVITY is still in its infancy. The advantages of space are that it is an almost weightless environment and a better vacuum than we can produce on Earth.

More experimenting than actual commercial manufacturing has gone on so far. The National Institute of Standards and Technology does sell tiny, perfectly spherical plastic beads (for calibrating instruments) that are made in space. Pharmaceutical companies have used a technique called electrophoresis to purify drugs. (Much of the cost of a drug comes in the purification process, which can sometimes be done more efficiently in space because gravity doesn't interfere.)

Semiconductors for the electronics industry must be exceptionally pure to function. Many impurities come from the container in which ingredients are processed. In the zero-g

Experimental drug processing in space   NASA

environment of space, materials can be processed without containers, levitating them with magnetic fields or sound waves to allow purer processing.

It may be possible to produce alloys in space from materials that won't mix on Earth because one floats on another, or to make light but strong foamed metals that have tiny bubbles evenly dispersed throughout the material.

Manned spacecraft in free fall are not totally weightless, as the crew moves around and small thrusters keep the craft in the proper orientation. Sometimes, too, waste products are vented into space. Therefore some materials processing will probably be done in separate, free-flying platforms that are visited by astronauts in order to renew supplies and remove finished materials.

It is probable that the greatest benefits from materials processing in microgravity will come from things we can't even imagine yet.

Perfect spheres made in zero g
NASA

Astronaut Guion Bluford prepares an experiment on the shuttle
NASA

MIR is the world's only operational space station. It follows the Soviet Salyut series first placed in space in 1971. The first 20-ton component of Mir, which means both "peace" and "world" in Russian, was launched February 20, 1986, aboard a Proton rocket from the Tyuratam spaceport. Its orbit is inclined about 52 degrees at an altitude of around 200 miles. Because of its size it is very bright and can easily be seen from Earth.

The station has been increased in size several times since launch. The original 43-foot-long, 14-foot-diameter, 21-ton station has had a 20-foot-long, 10.6-ton Kvant ("Quantum") astrophysics research module added on to it. The station has five docking ports, which allow both manned and unmanned resupply vessels to attach to the station. Three large solar panels supply electricity. Mir can accommodate a crew of up to six cosmonauts, but usually there are only two or three aboard except during crew changes. Several cosmonauts have spent over 200 days each in space aboard Mir. Robotic Progress tanker craft are launched and automatically rendezvous and dock with Mir to bring water, food, and supplies, and to retrieve materials from the experiments being conducted aboard. More additions are likely.

Among the activities carried out in Mir are astronomy observations, materials processing (including purification of drugs similar to the electrophoresis equipment carried aboard the space shuttle), manufacturing experimental alloys, Earth observations, studying the biomedical effects of zero g, and military research.

The large size and low orbit of Mir mean that it is subject to much atmospheric drag. Rocket engines aboard the Progress tankers are used to re-boost the station to higher altitudes.

Mir, the Soviet space station

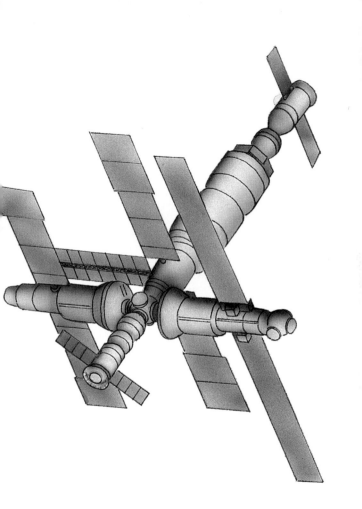

THE U.S. SPACE STATION, named "Freedom," will provide a permanently manned space research facility by the end of the 1990s. The basic design has one or more long beams with large solar panels and heat radiators to provide electricity. Several cylindrical modules—including one from Europe and one from Japan—will serve as a dormitory for crew, laboratories, and detachable logistics modules for storing food, fuel, and supplies. Canada will supply a mobile robot arm for use in construction around the station. The design could change.

By 2000 the station should be complete and capable of permanent manned occupancy. After that, a second phase of construction will add more large beams, solar-dynamic power supplies to increase the available electrical power, and more modules. A service bay will enable astronauts to construct spacecraft or to perform maintenance on spacecraft.

FREE-FLYING PLATFORMS near the main station will allow microgravity experiments. Possibly separate astronomy platforms will be built.

Later, a co-orbiting small station may be built by the European Space Agency and launched by the space shuttle for more microgravity work. Scientists will commute from the station to the platforms. A separate polar-orbiting platform is planned for Earth-observation missions; it will be smaller than the main station.

SPACETUGS, or space ferries, are being developed to carry payloads from one orbit to another, or from one spacecraft to another. Spacetugs may be automated or piloted. Further in the future we may have lunar shuttles to ferry crews and supplies between Earth orbit and a lunar base.

One concept of a U.S. space station   NASA

MILITARY USES OF SPACE

Most of the technology now used in civilian space programs came out of military research, just as did most aviation technology before it. Most rockets were originally designed to carry warheads. Many of the instruments used in astronomy and in Earth-observing satellites were designed for spy satellites (spysats). To national leaders concerned with their nations' defense, space is the "high ground" that must be used and occupied.

Other than antisatellite weapons and the rockets that carry conventional and nuclear warheads, much of the military use of space promotes world stability. The ability to see from space military movements and factories making military equipment, or even to eavesdrop on a potential enemy's communications, makes surprises and uninformed reactions less likely.

Some details are known about U.S. military satellites from unclassified sources. Most Soviet satellites are hidden under the catchall phrase "Kosmos," and much less is known about them by the public.

MILITARY COMMUNICATIONS SATELLITES allow worldwide coverage. Most U.S. long-distance military communication uses satellites. Since geostationary satellites are not useful from latitudes greater than 77 degrees (they would be too close to the horizon to get a good signal through the atmosphere), both major powers have non-geostationary commsats in orbits inclined about 63 degrees. The Soviet satellite system is called Molniya; the U.S. has the Satellite Data System. Such satellites allow nations to stay in touch with their naval and air forces in polar regions, and to relay data from spysats.

METEOROLOGY SATELLITES (metsats) provide military forces with important information on weather conditions worldwide. They can give data on clouds, icebergs, and precipitation, and they can track large, dangerous storms. This is important to troops and vessels, and it also helps target spy satellites to areas with little cloud cover. The U.S. military metsat system is called the Defense Meteorological Satellite Program.

NAVIGATION SATELLITES (navsats) provide military users such as ground troops, ships, aircraft, and missiles with very accurate position determination. The current U.S. system is called Navstar or Global Positioning System (see p. 137).

OCEAN SURVEILLANCE satellites track ships, and can even detect some submerged submarines. The U.S. currently has no such satellites. The Soviet Union has for many years used a series of Radar Ocean Reconnaissance Satellites.

MILITARY RESEARCH SATELLITES carry into space scientific experiments of interest to the defense forces, including work on new sensors, propulsion, and even sophisticated manufacturing techniques. Space shuttles are used both for military and civilian research missions; the space station may be as well.

EARLY WARNING SATELLITES provide national command forces with information on possible missile attacks and enemy military satellites. The current U.S. system is called the Defense Support Program, a series of three geostationary satellites stationed over the Atlantic, Indian, and Pacific oceans watching for ballistic missile launches from land bases or from submarines.

NUCLEAR DETECTION SATELLITES watch from orbit for the intense flashes of nuclear bombs as a way of enforcing test bans. The earliest U.S. detection satellites were called Vela. Now that function is also performed by the Integrated Operational Nuclear Detection System carried piggyback on Navstar satellites.

ELECTRONIC INTELLIGENCE SATELLITES are some of the most secret satellites used by defense organizations. Often called ferret satellites, they are used to eavesdrop on radio and radar transmissions and can reportedly even pick up, from low orbit, telephone calls carried on terrestrial microwave links. Others are in geostationary orbits.

The first U.S. series was called Rhyolite. More advanced satellites reportedly have code names like Aquacade, Magnum, Chalet, and Jumpseat.

Military navigation satellite
NASA

Soviet Radar Ocean Surveillance Satellite

SPY SATELLITES (spysats) are also highly secret. They are placed in low orbits, and have maneuvering engines that allow them to swoop even lower for closer photographs, to return to higher orbits, and to change orbits in order to scan different target areas.

Following the early U.S. Discoverer series, the "Big Bird" satellites, also known as KH("Keyhole")-9, were launched between 1971 and 1984. These huge spysats were 50 feet long, 10 feet in diameter, and weighed 25,000 pounds. They contained long-focal-length cameras. (They also sometimes contained some electronic eavesdropping capabilities.) Photographs were developed in the spacecraft automatically. Some were scanned electronically and the pictures sent back to receiving stations on Earth by radio. For the highest-resolution pictures, the film was transferred to one of six pods that were then jettisoned and returned to Earth by parachute, being caught in midair by airplanes.

Unofficial sources claim Big Birds could recognize objects as small as a foot across from 100 miles up. A standing joke is that the CIA has nicknames for every Russian truck driver. Big Birds were exploded in orbit or commanded to reenter and burn up at the end of their operational lives to prevent them from falling into foreign hands.

The Titan-3 launcher was specifically developed to launch the Big Birds. Current spysats are launched by newer Titan models or by the space shuttle. Big Bird used the Agena upper stage, which remained attached to it to enable it to maneuver over different geographical areas of interest.

The newer U.S. spysats, beginning in the mid-1980s, are the KH-11 and KH-12, supposedly with ever-better resolution and longer operational lifetimes in orbit. Technology has improved, so film return is no longer necessary.

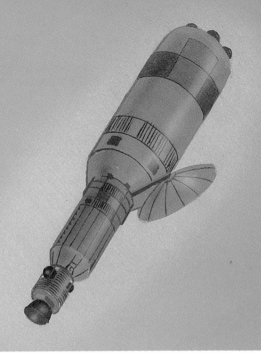

Artist's conception a Big Bird spysat

LACROSSE is the name of a new radar spysat first launched aboard a space shuttle in 1988. It is the first U.S. active radar satellite, capable of making images of what it "sees" with radio waves. The radar is capable of peering through clouds and of being used at night as well.

Lacrosse satellites are about 150 feet across, and are placed in orbits about 400 miles up. Unclassified sources claim they can resolve objects as small as about 3 feet across. Their connection with the new KH-12 spysats is not known outside the military.

FUTURE SPACE MISSIONS

So far our space missions have been confined to the solar system, and even that we have explored only spottily. Only the Moon has been visited by humans in person.

We are technically limited in our explorations by the power of our rockets, by our ability to carry food and fuel, and by one apparently insuperable physical limitation: the speed of light. No object or signal we now know of can travel faster than 186,000 miles per second.

Further exploration and exploitation of space will depend upon more powerful and efficient rocket engines. Among the possibilities are ion engines, which, although they produce only a small thrust, can provide power for a long time and make very efficient use of fuel.

Other low-thrust techniques such as solar sails may be used. Technologies such as fusion ramjets are far in the future. Such fanciful ideas as anti-gravity, "warp drive," and so on are likely to remain in the realm of science fiction, although we should never underestimate the potential for research breakthroughs.

Much will be done robotically, sending "smart" machines into distant and dangerous environments. Robots will become increasingly autonomous. This is especially important the farther from Earth our probes go, and thus the longer it takes to communicate with them via radio.

The next few pages outline several possible future space activities, some almost certain to occur within the next couple of decades, some that will perhaps never occur.

Whether any particular project happens is not so important as the fact that, just as the 20th century saw terrestrials take the first tentative steps off our planet after 5 billion years of evolution, the 21st century will see the permanent expansion of human beings into the solar system and the universe beyond.

A future base on Mars might look like this

LUNAR BASES will be among the next manned space activities. The Moon could be mined for materials to be used in building space settlements and spacecraft, and for fuel. The Moon is a good site for astronomical radio and optical observatories. Lunar orbit may be a good place to assemble huge spacecraft for planetary missions.

A MANNED MISSION TO MARS is among the programs being considered by the U.S. and the Soviet Union, perhaps as an international venture that includes Europe and Japan.

Because the Moon and Mars have no or little protective atmosphere or magnetic field, bases there would have to be dug several feet below the surface to avoid cosmic rays and meteoroids. Humans would need to wear pressure suits for work on the surface. Within a couple of decades there may even be tourist visits to the Moon.

The first true extraterrestrial human is likely to be born in a lunar base within the next couple of decades.

ASTEROID MINES may become practical in the future. Although the asteroid belt between Mars and Jupiter seems very distant, it takes only about the same amount of energy to bring a pound of asteroidal material from there to a near-Earth orbit as it does to bring a pound of material from Earth's surface to orbit up through the Earth's strong gravitational field.

A small nickel-iron asteroid, one mile in diameter, contains about a decade's worth of the total output of all iron smelters in the entire world, and it is already refined to about 95 percent purity! The silicon, oxygen, and other elements in the rocky asteroids would be valuable on the Moon, on Mars, or in space settlements.

It would take several years to ferry asteroidal material economically back to near-Earth, perhaps using solar sails or ion rockets utilizing the asteroid itself as fuel. A few asteroids pass close to Earth, and we may be able to rendezvous with them.

Mining an asteroid

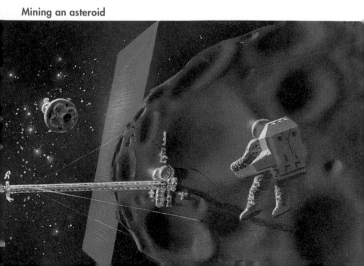

Concept for a space colony   Inside a space colony

SPACE SETTLEMENTS, or space colonies, will be huge "cities" built in space using mostly extraterrestrial materials. The most likely location would be in the Moon's orbit, 60 degrees ahead of or behind it, at the locations called by space scientists L4 and L5.

These settlements would be spheres or cylinders miles across in size, with fully self-contained, self-sustaining ecosystems, including houses, animals, plants, lakes, agriculture, and everything else needed to support thousands of people indefinitely. Day and night cycles would be provided by rotating the settlement and by opening and closing shutters. Rotation would also provide a kind of "artificial gravity."

The space settlers may be employed in research, manufacture of spacecraft, refining lunar materials, and many other activities. Some may come to retire in the reduced "gravity" of the settlement. Others may come to vacation. While such settlements are at least several decades away, there are now many people who would like to live in one.

SOLAR SAILS, or light sails, use the power of sunlight to move a spacecraft. Only Earth-bound experimental models have been built so far, but one day they may be used to transport cargo across vast distances. Although their thrust is extremely low, they need carry no fuel and no engines.

Sunlight exerts a slight pressure, the same force that makes a comet's gaseous tail stream away from the sun. A large lightweight sail, miles across, made of aluminized plastic, could use this light pressure to provide a few pounds of force. They would be controlled much as a sailboat on a lake is, by altering the position of the sail to slow down or speed up as needed. At distances that are much farther from the Sun than the asteroid belt, there would be too little sunlight to work well.

Such a technique could be used to ferry materials from asteroids to near-Earth. Although the journey would take many years, it would be inexpensive. There are plans for a solar-sail trial race in the 1990s.

Solar sailing

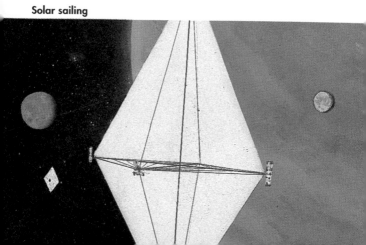

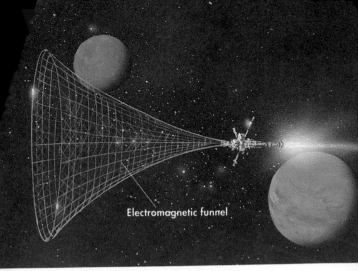

Electromagnetic funnel

A fusion ramjet starship

INTERSTELLAR FLIGHT is a dream out of reach for some time. Even at the speed of light it would take 4.3 years to reach the nearest star system. It now seems impossible to travel faster than light.

The Hubble Space Telescope may tell us if any of our neighboring stars have planets around them. Robotic probes could journey there, taking decades and sending findings back by radio or laser beams. Manned starflight is much further in the future, and flights would probably take so long they would be one-way voyages.

One technology imagined for such a futuristic mission is the fusion ramjet, which generates a miles-wide funnel-shaped magnetic field in front of it. Its motion through the thin interstellar gas would capture hydrogen atoms, which would then be compressed and expelled out the rear. As yet no one knows how to make such a craft. Still, it is exciting to consider that someday humans may journey among the stars.

ORGANIZATIONS AND RESOURCES

SPACE ORGANIZATIONS can give you information on what is happening in the space program. They usually publish informative magazines and hold meetings, and they may actively promote space activities. The major ones are:

National Space Society
922 Pennsylvania Avenue, SE
Washington, DC 20003

British Interplanetary Society
27/29 South Lambeth Road
London SW8 1SZ, England

Planetary Society
65 North Catalina Avenue
Pasadena, CA 91106

NASA FACILITIES often have public visitor centers, and some offer special resources for teachers. If you live or visit near one of these, contact them ahead of time for more information.

Visitor Information Center
Goddard Space Flight Center
Greenbelt, MD 20771

Public Affairs Office
Lewis Research Center
Cleveland, OH 44135

Public Affairs Office
Johnson Space Center
Houston, TX 77058

Public Affairs Office
Ames Research Center
Mountainview, CA 94035

Visitor Information Center
Kennedy Space Center
Merritt Island, FL 32899

Public Affairs Office
Jet Propulsion Laboratory
Pasadena, CA 91103

MAJOR SPACE MUSEUMS AND EXHIBITS:

National Air and Space Museum
6th and Independence Ave., SW
Washington, DC 20560

Alabama Space and Rocket Center
1 Tranquility Base
Huntsville, AL 35807

BOOKS AND MAGAZINES
Following are recommended books about a variety of space activities:

Blonstein, Larry. *Communications Satellites*. John Wiley, New York, 1987.

Chochran, Curtis D., Dennis M. Gorman, and Joseph D. Dumoulin, eds. *Space Handbook*. Air University Press Publication AU-18, U.S. Government Printing Office, Washington, DC, 1985.

Gatland, Kenneth. *The Illustrated Encyclopedia of Space Technology*. Harmony Books, New York, 1981.

Goldman, Nathan C. *Space Commerce*. Ballinger Publishing Company, Cambridge, MA, 1985.

Hobbs, David. *An Illustrated Guide to Space Warfare*. Salamander Books, New York, 1986.

Sheffield, Charles. *Earthwatch: A Survey of the World from Space*. Macmillan Publishing Co., Inc., New York, 1981.

Turnill, Reginald, ed. *Jane's Spaceflight Directory for 1987*. Jane's Publishing Inc., 115 Fifth Avenue, New York, 1987.

Von Braun, Wernher, Frederick I. Ordway, David Dooling, and Fred C. Durant. *Space Travel: A History*. Harper & Row, New York, 1985.

Space events happen faster than books can keep up with them. Magazines are indispensable for keeping up-to-date. These are the major ones:

Ad Astra
National Space Society
922 Pennsylvania Avenue, SE
Washington, DC 20003

Spaceflight
British Interplanetary Society
27/29 South Lambeth Road
London SW8 1SZ, England

Aviation Week & Space Technology
P. O. Box 1505
Neptune, NJ 07753

INDEX